Hungary and NATO

Hungary and NATO

Problems in Civil-Military Relations

Jeffrey Simon

ROWMAN & LITTLEFIELD PUBLISHERS, INC.
Lanham • Boulder • New York • Toronto • Oxford

ROWMAN & LITTLEFIELD PUBLISHERS, INC.

Published in the United States of America
by Rowman & Littlefield Publishers, Inc.
A wholly owned subsidiary of of The Rowman & Littlefield Publishing Group
4501 Forbes Boulevard, Suite 200, Lanham, Maryland 20706
www.rowmanlittlefield.com

P.O. Box 317, Oxford OX2 9RU, United Kingdom

British Library Cataloguing in Publication Information Available

Library of Congress Cataloging-in-Publication Data

Simon, Jeffrey, 1942–
   Hungary and NATO : problems in civil-military relations / Jeffrey Simon.
      p. cm.
Includes bibliographical references and index.
   ISBN 0-7425-2850-2 (cloth : alk. paper) — ISBN 0-7425-2851-0 (pbk. : alk. paper)
   1. Hungary—Politics and government—1989– 2. Civil supremacy over the military—
Hungary. 3. Civil military relations—Hungary. 4. North Atlantic Treaty Organization—
Hungary. I. Title.
   DB958.3 .S56 2003
   320.9439'09'049—dc21

                                                                    2003008109

Printed in the United States of America

♾™ The paper used in this publication meets the minimum requirements of American
National Standard for Information Sciences—Permanence of Paper for Printed Library
Materials, ANSI/NISO Z39.48-1992.

Dedicated in memory of my mother and
father, who are together again

lat. 26 12 272
long. 80 04 526

Contents

Acronyms

AFD	Alliance of Free Democrats
AFSOUTH	Allied Forces Southern Europe
APC	Armored Personnel Carrier
APV	State Privatization and Asset Corporation
ASOC	Airspace Sovereignty Operations Center
BG	Brigadier General
CFE	Conventional Forces in Europe Treaty
CMX	Crisis Management Exercise
ColGen	Colonel General
CoS	Chief of Staff
DASA	DaimlerChrysler Aerospace
DPQ	Defense Planning Questionnaire
EU	European Union
Fidesz	Alliance of Young Democrats
Fidesz-MPP	Federation of Young Democrats-Hungarian Civic Party
FKGP	Smallholders Party
Ft	Forint currency unit
GDP	Gross Domestic Product
HDF	Hungarian Defenders of the Homeland
HSP	Hungarian Socialist Party
HSWP	Hungarian Socialist Worker's Party
IFF	Identify Friend or Foe
IFOR	Bosnia Implementation Force
IH	Information Office
ISAF	International Security Assistance Force
KBH	Military Security Office

KDNP	Christian Democratic People's Party
KFH	Military Intelligence Office
KFOR	Kosovo Force
LTG	Lieutenant General
MBT	Main Battle Tank
MDF	Hungarian Democratic Forum
MG	Major General
MIEP	Hungarian Life and Justice Party
MP	Member of Parliament
NAC	North Atlantic Council
NACC	North Atlantic Cooperation Council
NATO	North Atlantic Treaty Organization
NBH	National Security Office
NCO	Noncommissioned Officer
NGSF	Northern Group of Soviet Forces
OEF	Operation Enduring Freedom
OSCE	Organization for Security and Cooperation in Europe
PARP	Planning and Review Program
PFP	Partnership for Peace
PPBS	Planning, Programming, and Budgeting System
SACEUR	Supreme Allied Commander Europe
SFOR	Bosnia Stabilization Force
SGSF	Southern Group of Soviet Forces
SHAPE	Supreme Headquarters Allied Powers Europe
SOFA	Status of Forces Agreement
STANAG	Standing Agreement [NATO]
UN	United Nations
UNFICYP	United Nations Peacekeeping Force in Cyprus
US	United States of America
WTO	Warsaw Treaty Organization

Acknowledgments

Since the revolutions of 1989–1990, I have had the privilege and opportunity to work with many counterparts in Central and Eastern Europe on a continuing basis. Working with these colleagues in an effort to implement real defense reforms and to develop democratic oversight of the military has been not only rewarding and educational, but has also led to some extremely close friendships. The three books in this series are an outgrowth of that experience and I must acknowledge the assistance and support of the numerous friends in Central Europe, NATO, and the U.S. government who have shared their knowledge and advice and read and critiqued portions of the manuscript. They are too numerous to name, but they know who they are and this book could not have been written without them.

I have also had the great fortune to work at the National Defense University's Institute for National Strategic Studies, which provided me with the unique opportunity to work in my chosen field of Central and Eastern Europe during a critical historical juncture. During this period, I have enjoyed the support of university presidents Vice Admiral J. A. Baldwin, LTG Paul G. Cerjan, LTG Ervin J. Rokke, LTG Richard A. Chilcoat, and Vice Admiral Paul G. Gaffney and have been extremely lucky to serve three extremely capable institute directors—Alvin H. Bernstein, Hans A. Binnendijk, and Stephen J. Flanagan—who have provided enormous support and continuous encouragement in these efforts. In addition, I would like to acknowledge the support of Jeffrey D. Smotherman of the Institute's Publication Directorate and my intern Joyce Hatfield, and to thank the National Defense University Foundation for its support of this project on civil-military relations. In addition, I would like to thank Matt Hammon and April Leo of Rowman & Littlefield for the enthusiasm and ideas they provided during this project.

Finally, and most important, I want to acknowledge the enormous patience, understanding, and encouragement provided by my wife, Katherine, and four wonderful children—Lauren Meiling, Adam Lee, Jean Paul, and Tamara Danielle.

NATO Membership

NATO Members not shown on map
- United States
- Canada

Country Abbreviations:
Belg. = Belgium
Cr. = Croatia
L. = Lichtenstein
Lux. = Luxembourg
Mac. = Macedonia
Neth. = Netherlands
Sl. = Slovenia
Slov. = Slovak Republic
Switz. = Switzerland

City Abbreviations:
Br. = Bratislava   Sa. Sarajevo
L. = Ljubljana   T. = Tirane
S. = Skopje   Z. = Zagreb

0   500 Miles
0   500 Kilometers

NATO members at the end of the Cold War

Members admitted to NATO in 1999

Countries anticipating NATO membership in 2002

⊛ Capitals (selected)

Countries aspiring to NATO membership

Countries with strategic relationship to NATO

* East Germany joins NATO with the reunification of Germany in 1990

1

1989 Revolution and Democratic Control of the Military

When revolution found Central and Eastern Europe in 1989–1990, it came in many forms, each of which had a different initial influence on the democratic transition. Felipe Aguero has called the first form transition through regime defeat, where the authoritarian regime collapsed and the ruling elites had little choice but to relinquish office. The regimes in the Czechoslovak Socialist Republic (CSSR) and the former German Democratic Republic (GDR) failed and had little control over the agenda or timing of the transition. The second form is transition through transaction. In this case the authoritarian government initiated the process of liberalization and remained a decisive actor through the transition. Romania fits this model. The third involves transition through extrication. Here the authoritarian government was able to negotiate crucial features of the transition, although in a position of less strength than transition through transaction. Poland and Hungary fit this case.[1]

Hungary's democratic transformation, in which the authoritarian government was able to negotiate critical features of the transition, was more similar to Poland's multistage evolutionary transition than to the sudden revolutions resulting from regime defeat in East Germany and Czechoslovakia. Hungarian reform differed from Czechoslovakian in that it lacked a politician like Vaclav Havel who had gained the confidence of society through long years of shared battles. Polish reform had a leader in Lech Walesa with strong indigenous institutions—the Church and Solidarity—from the outside with the effective cooperation of a corrupt party apparatus. The absence of a well-known strong leader would make it more difficult for Hungarian leaders to establish their legitimacy.[2]

In Hungary, with the Hungarian Socialist Worker's Party (HSWP) influence waning as a result of years of economic degeneration and deepening

1

crisis within the party apparatus, communist reformers inside the system engineered Hungary's reform in league with outside opposition. When the previously disunited Hungarian opposition reconciled its differences, the HSWP reform leadership committed itself to make the transition to a democratic multiparty system.

A major step toward coming to grips with Hungary's communist past occurred on 16 June 1989 when over 100,000 people took part in public funeral services for Imre Nagy who had been executed three decades earlier. Over the summer Hungary began to dismantle the "iron curtain" on its western border, and in September it opened the border for East German refugees to travel to the Federal Republic of Germany. This action sparked the exodus of East Germans to the west, and ignited the revolution in East Germany, creating a domino effect in Czechoslovakia.

During the summer and early fall, the HSWP's round-table negotiations with eight opposition parties resulted in an 18 September accord that cleared the way for multiparty elections. At the 6–9 October 1989 HSWP congress Imre Pozsgay's reform wing of the HSWP transformed the discredited party by changing its name to the Hungarian Socialist Party (HSP). It adopted a progressive program that embraced multiparty parliamentary democracy, respect for civil liberties, and a mixed economy.

Most observers anticipated that the March 1990 parliamentary elections would radically reduce communist HSP representation and result in a coalition government, including a number of opposition parties. Expecting this, Prime Minister Miklos Nemeth, in an effort to insulate the military from the forthcoming political change, announced on 1 December 1989 a major defense ministry and military reform, which included changing the military command.

The two major winners—the Hungarian Democratic Forum (MDF) with 165 of parliament's 386 seats and Alliance of Free Democrats (AFD) with 92 seats—formed a pact in late April and a new noncommunist government was sworn into power on 23 May 1990. Under terms of the agreement, the AFD agreed to reduce the amount of legislation requiring two-thirds majority vote in exchange for electing Arpad Goncz (AFD) as president.[3] Jozsef Antall, a member of the MDF, became prime minister. The first stage of Hungary's revolution was completed in August 1990, when parliament formally elected interim president Arpad Goncz president.

The second stage of Hungary's revolution commenced as relations between the MDF and AFD parties worsened from the end of 1990 through 1991. Significant differences developed over spheres of authority between the prime minister and president; these were challenged and resolved after the government's successful Constitutional Court challenge at the end of

1991. In 1992 a new defense reform was implemented to redress the effects of Miklos Nemeth's 1 December 1989 reform; and the MDF began to tighten its political control over the defense ministry and other key government institutions.

The third stage began with the May 1994 parliamentary elections that returned the Hungarian Socialist Party (HSP) to power. The new challenge, which remained unfulfilled, became the need to write a constitution that all the Hungarian electorate would consider legitimate. Although efforts to unify the General Staff with the defense ministry and civilianize the ministry of defense were also slow in evolving, Hungary did receive an invitation to join NATO in July 1997.

The fourth stage began in June 1998 when the parliamentary elections returned a center-right government led by the Young Democrats (Fidesz) in coalition with the MDF and Smallholders parties. Their challenge would be to complete NATO integration, continue the EU process, and consolidate the defense and military reform. When their mandate ended and the HSP returned to power in May 2002, it had become apparent that Hungary was unable to meet its Alliance obligations. Hungary's failure was due, in large part, to a serious dysfunction in civil-military relations and dramatizes the need to provide effective democratic oversight of the military.

WHAT CONSTITUTES DEMOCRATIC OVERSIGHT OF THE MILITARY?

Effective democratic oversight of the military can be judged by the following four factors:[4]

1. *A clear division of authority between president and the government (prime minister and defense/interior minister) in constitutions or through public law.* The law should clearly establish who commands and controls the military and promotes military officers in peacetime, who holds emergency powers in crisis, and who has authority to make the transition to war. Underlining these formalities is evidence of the spirit of tolerance and respect for legitimacy between president and prime minister (government) who may often be from different political parties or ideological persuasion.
2. *Parliamentary oversight of the military through control of the defense budget and defense committee.* Parliament's role in deploying armed forces in peacetime, emergency, and war must be clear. Underlining

these formalities is the need for the Defense and Security and Foreign Affairs Committees to provide minority and opposition parties with transparent information and allow consultation particularly on normal policy issues such as defense budgets, defense plans, and monitoring of the armed forces and on extraordinary commissions investigating defense/security violations. Parliamentary committees need staff expertise and adequate information to provide oversight and effective liaison with defense and interior ministries to help develop bipartisan consensus on defense and security matters. Similarly intelligence oversight committees should provide access to opposition parties.

3. *Peacetime government oversight of General Staffs and military commanders through civilian defense ministries.* Defense ministry management should include preparation of the defense budget, access to intelligence, strategic planning, and defense planning to include force structure development and deployment, arms acquisitions and property management, and personnel policies to include career development and military promotions. The defense ministry also needs press and legislative liaison capabilities to communicate to the public and parliament how and why public resources are being expended.

4. *Restoration of military prestige, trustworthiness, and accountability for the armed forces to be effective.* Because the military was controlled by the Soviet High Command through the Warsaw Pact (and top-secret statute system) and often used as an instrument of external or internal oppression during communist rule, postcommunist General Staffs must be apolitical and professional, and society needs to perceive the military as being socially useful, militarily capable, and under effective civilian control. In addition to the necessary constitutional and institutional arrangements, there needs to be a Legal Framework and Code of Conduct for professional soldiers and conscript citizens that would allow them to disobey illegal orders. Military training levels and equipment must also be sufficient to allow a necessary level of professional prestige. This requires social support for conscription (for the foreseeable future) and a predictable stream of material resources (adequate defense budgets) that the defense ministry can "sell" to parliament and the broader society. In 2002, the Central European reality was different. Most militaries retained 25–35 percent of their 1988 manpower levels and 25–35 percent of the defense budgets in real terms. In sum, their readiness, training, and modernization levels have deteriorated significantly; in

some cases, even to raise questions about their capacity to defend their territory let alone to participate in coalition defense tasks. Finally, though the Central European militaries have evidenced significant reforms and been restructured to accommodate NATO, force modernization continues to be greatly restrained by scarce resources.

Hungary and its respective governments since 1989–1990 have been grappling with the challenge of developing democratic oversight of the military and integrating into NATO. In large degree, its lack of success thus far has been the result of a serious dysfunctional relationship between society and the military and will not likely be changed until this dysfunction is rectified.

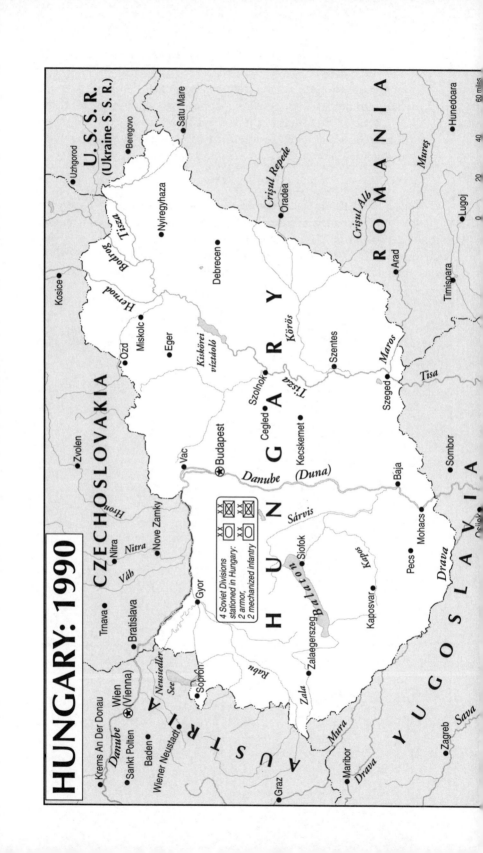

HUNGARY: 1990

CZECHOSLOVAKIA

U.S.S.R.
(Ukraine S. S. R.)

ROMANIA

HUNGARY

AUSTRIA

YUGOSLAVIA

4 Soviet Divisions
stationed in Hungary:
2 armor,
2 mechanized infantry

Krems An Der Donau
Sankt Polten
Baden
Wiener Neustadt
Wien (Vienna)
Graz
Maribor
Zagreb

Zvolen
Kosice
Trnava
Bratislava
Nitra
Nove Zamky
Sopron
Gyor
Zalaegerszeg
Kaposvar
Siofok
Pecs
Mohacs

Uzhgorod
Beregovo
Satu Mare
Nyiregyhaza
Debrecen
Miskolc
Ozd
Eger
Szolnok
Budapest
Vac
Cegled
Kecskemet
Baja
Szentes
Szeged
Arad
Oradea
Timisoara
Lugoj
Hunedoara
Sombor

Danube (Duna)
Tisza
Bodrog
Hernod
Körös
Maros
Tisa
Sárvis
Kapos
Balaton
Zala
Mura
Drava
Sava
Crisul Repede
Crisul Alb
Mures
Neusiedler See
Raba
Váh
Hron
Nitra

60 miles
0 20 40

2

Miklos Nemeth's Legacy
and the Need for Defense Reform

The purpose of Hungary's defense reform has been to establish democratic (Parliament and government) oversight of the defense ministry and Hungarian People's Army (now Magyar Honvedseg) or Hungarian Defenders of the Homeland (HDF). Hungary also had to clarify the lines of authority between the president and government (prime minister and civilian defense minister) in peacetime and in wartime, and had to establish defense ministry oversight and management of the General Staff. Finally, the reform had to remove Soviet and HSWP influence from the military establishment, to ensure that Hungarian military forces were sufficient to guarantee the integrity of Hungary, and to return the armed forces to Hungarian society.

Under the communist system, Hungarian national security policy (as in Poland) was formulated by a small group headed by the HSWP first secretary in his capacity as president of the Defense Council, and in the HSWP Central Committee by the secretary in charge of national defense, with perhaps the addition of the commander-in-chief of the armed forces and the defense minister who was a military officer. In July 1989, when Hungary was still under the Warsaw Pact and four Soviet divisions—the Southern Group of Soviet Forces (SGSF)—remained in the country, Prime Minister Miklos Nemeth noted that Hungary's new national defense policy needed to make clear that Hungary's national armed forces were in the hands of democratic power under appropriate and strict control. When asked who was commander-in-chief of the armed forces, Nemeth answered "it is not possible at present to give an unequivocal reply to this."[1]

During 1989 the communist-dominated Ministry of Justice drafted an entirely new Hungarian constitution (to succeed the 1949 communist constitution), based upon the principles articulated at the spring round-table talks. The parliament that passed that constitution in October 1989 was still dominated by members of the communist party.

According to the October 1989 constitutional changes, National Assembly representatives are elected for four-year terms, and the president, who is elected by the National Assembly for five years, is commander-in-chief of the armed forces.[2] Only Parliament is entitled to make decisions concerning the use of the armed forces.[3] According to Article 19 of the Constitution, the National Assembly has the power to declare a state of war and conclusion of peace. In the event of war, it declares a state of emergency and sets up the Defense Council. If the National Assembly is unable to convene, the president assumes these powers.

When so empowered, the Defense Council—chaired by the president[4]— has the authority to deploy armed forces abroad and within the country.[5] During peacetime, the prime minister, elected by a majority of the National Assembly, and ministers of the government "control the operation of the armed forces, the police, and other organs of policing."[6]

On 1 December 1989 Hungary's defense reform divided the defense ministry into two separate entities: a defense ministry subordinate to the prime minister and a commander of the Hungarian army subordinate to the president (see figure 2.1).[7] When the defense reform was announced, the Nemeth government's intention was to take the armed forces, which until then were under direct party command, and remove them from the direct influence of the future noncommunist government, which was expected to exercise power following the then anticipated March 1990 multiparty elections.

The new reform was clearly intended to put the armed forces under communist control by removing the core of the army cadres from the defense ministry and placing them under a commander of the Hungarian army subordinate to the president. As a result of the reform, the president—who at the time was assumed would be communist-reformer Imre Pozsgay— became the commander-in-chief of the army. Whereas in most other parliamentary systems a clear line of authority exists from prime minister to defense minister to chief of staff (CoS), after the December 1989 Hungarian defense reform the line of authority went directly from the president to Commander of the Hungarian Army to the CoS, with the government (meaning the ministry of defense) basically out of the chain of command.

One unintended and unfortunate result of the reform was increased tension between the president and the government (prime minister and defense minister). Subordinate to the prime minister (before elections Miklos Nemeth, Jozsef Antall of the MDF after) and Council of Ministers is the defense minister (then Ferenc Karpati, after May 1990 an MDF civilian Lajos

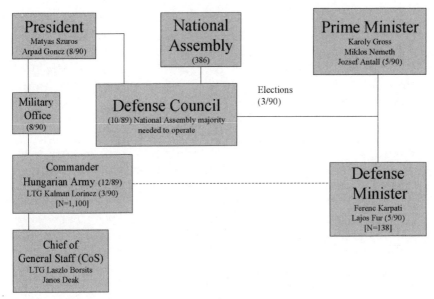

Figure 2.1.   Nemeth/Karpati Defense Reform, 1989–1990

Fur) who maintained a relatively small staff (the total defense ministry comprised 138 people) and was responsible for state administration tasks and military policy. After the 1989 defense reform the defense ministry dealt more with social and political questions, matters that parliament normally dealt with.[8]

After the 1989 defense reform, the army remained subordinate to the president of the republic (then Matyas Szuros, since August 1990 Arpad Goncz of the AFD), and control over the army as of March 1990 was now exercised by a new (1,100-man) commander of the Hungarian army (LTG Kalman Lorincz) who, as commander-in-chief of the armed forces, supervised actual military tasks.[9] Under the defense reform, the president also has authority to appoint generals.[10] According to the LTG Chief of General Staff Laszlo Borsits:

> The higher military leadership is exercised by *the Commander-in-Chief, through the General Staff.* The troops are directly commanded by the field army corps staff and the home air defense corps staff. On the operational-tactical level of command the corps, brigade, and battalion staffs perform the task of leadership.[11]

Hungarian concerns about control of Hungarian forces during an emergency and authority to make the transition to war were evident in the October 1989 National Assembly debate over the new draft constitution. At the time, only a "qualified majority . . . in the National Assembly"[12] could declare a state of emergency or war, which brought into being the Defense Council in order to exert extraordinary measures.

Subsequent National Assembly Defense Committee sessions focused on the issue of the Soviet Union's control of Hungarian armed forces, in part, because four Soviet divisions (SGSF) remained in the country as a result of Hungary's membership in the Warsaw Treaty Organization. Parliamentary discussions evinced sensitivity about the illegality of the 1968 invasion of Czechoslovakia and in the problems of HSWP control over the army. Hungary's participation in the 1968 Warsaw Pact invasion focused attention on the issue of command and control of armed forces. Bela Biszku, who had been HSWP Central Committee secretary between 1962 and 1978, told the National Assembly's Defense Committee on 3 January 1990 that the related command for intervention in 1968 was "most certainly" given to the Hungarian defense minister by the Warsaw Pact's Combined Armed Forces commander-in-chief.[13] In response, the National Assembly amended the defense law during February 1990 so as to grant itself the authority to decide on the deployment of Hungarian armed forces abroad or in Hungary.[14] After the Defense Committee blamed the HSWP for the illegal 1968 invasion, it concluded on 2 March 1990 "that party direction of the army must in all events be abolished."[15]

In summary, Miklos Nemeth's legacy was to negotiate critical features of the transition and set the stage for Hungarian postcommunist defense reform efforts to achieve democratic oversight of the armed forces. Its passage of the October 1989 Constitution and December 1989 defense reform established the problem of needing to determine *who* was "commander" of the armed forces—the president or the government. It did, though, set the stage for parliamentary efforts to establish rudimentary oversight of the forces, since the SGSF remained in Hungary until June 1991, by requiring a qualified majority of parliament to declare a state of emergency. The General Staff, though, was separated from the defense ministry and remained shielded from governmental oversight.

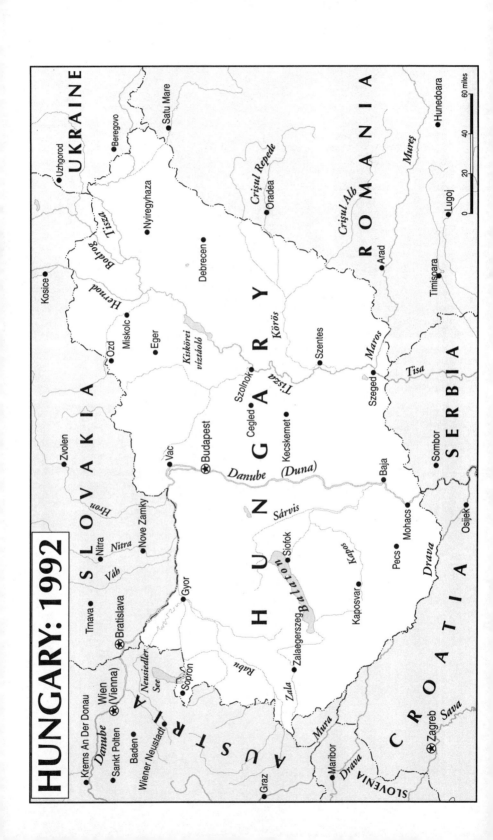

HUNGARY: 1992

3

1990 Parliamentary Elections: The Rise of Jozsef Antall

After the first free parliamentary elections were held on 25 March and 8 April 1990,[1] six parties, of the twenty-eight parties fielding candidates, passed the 4 percent threshold and entered the 386-seat Parliament (see table 3.1 on p. 14). In the first round 65 percent of the voters participated; and in the second round only 45.5 percent cast ballots. The Hungarian Democratic Forum (MDF) with 165 of Parliament's 386 seats formed a coalition with the Smallholders Party (FKGP) with 44 seats and the Christian Democratic People's Party (KDNP) with 21 seats. The parliamentary opposition consisted of the Hungarian Socialist Party (HSP), Alliance of Young Democrats (Fidesz), and Alliance of Free Democrats (AFD).

After the March 1990 elections, the governing MDF coalition and opposition Alliance of Free Democrats agreed to many significant amendments to the new constitution. The National Assembly amended the Constitution on 19 June 1990 to change some of the more objectionable provisions of the former communist government that related to the use of force. For example, Chapter VIII, which deals with "The Armed Forces and Police," now specifically required a *two-thirds* (rather than simple) majority of the National Assembly to employ these forces, thereby ensuring parliamentary control over them.[2]

Another significant defense reform involved intelligence. The Council of Ministers established four offices (two civilian and two military) to deal with intelligence: The National Security Office (NBH) under MG Sandor Simon (who replaced Colonel Lajos Nagy) and the Information Office (IH) under MG Kalman Kocsis (who replaced Colonel Istvan Dercze) have nationwide responsibility, are under independent jurisdiction, and are overseen by civilian minister without portfolio Andras Galszecsy, who receives directions through the Office of the Prime Minister. The third and fourth, the Military Security Office (KBH) under MG Karoly Gyaraki (who replaced

Table 3.1.   Hungarian Parliamentary Elections: 25 March and 8 April 1990

Political Party	Seats	% of Votes
Government (N = 230)		
Hungarian Democratic Forum (MDF)	165	24.7
Independent Smallholders and Citizens Party (FKGP)	44	11.7
Christian Democratic People's Party (KDNP)	21	6.8
Opposition (N = 146)		
Alliance of Free Democrats (AFD)	92	21.3
Hungarian Socialist Party (HSP)	33	10.9
Alliance of Young Democrats (Fidesz)	21	8.9
Independent and Joint Candidates	10	N.A.
Total	386	100.0

Colonel Jeno Gubicza) and Military Intelligence Office (KFH) under MG Janos Kovacs are part of the Hungarian Honved Forces, funded through the defense budget, and overseen by Defense Minister Lajos Fur.[3]

On 14 February 1990, the military security function was transferred to the Hungarian Defense Forces from the Interior Ministry III/IV group command and became an independent organization. Its functions are to protect against foreign intelligence activities, prevent insurrections and danger to military preparedness, provide physical security for military facilities, and protect persons performing confidential functions.[4] Both the Military Intelligence Office and the civilian Information Office have global responsibility, while the National Security Office and Military Security Office are confined to operating in Hungary.

While the goal of the general defense reform amounted to reassuming national control of the Hungarian military from the Soviet Union since Soviet forces were still in Hungary, the defense reform also created new problems between presidential and governmental authority. On 3 August 1990 Parliament elected Arpad Goncz (AFD) president.[5] In order to fulfill his constitutional duties to approve Hungary's armed forces defense plan and to declare an emergency or convene the Defense Council in case the National Assembly is impeded from doing so, President Goncz created a Military Office to liaise with the Commander of the Hungarian Defense Forces. MG Robert Pick, who headed and managed the activities of the office until 1995, informed President Goncz on subjects related to general military policy and military diplomacy and acted as the core staff of the commander-in-chief during the transition period between peace and war.[6]

Though the December 1989 reform was successfully implemented, intervening events during 1990—such as parliamentary elections resulting in a six-party coalition government producing a prime minister and president from different political parties—and the 1991 collapse of the Warsaw Pact and withdrawal of Soviet forces from Hungary—created new civil-military problems. In fact one might argue that the 1989 defense reform created more problems than it solved.

The 1989 defense reform contributed to confusion and differences of opinion over the span of authority between the commander of the Hungarian army and the defense minister. Though Lajos Fur replaced Ferenc Karpati on 23 May 1990 and became Hungary's first civilian defense minister after four decades of communist rule,[7] by September Defense Minister Fur was expressing concern about limits to his authority. He believed that officer training institutions, the Institute of Military History, and the management of all cultural areas, which were under military authority, should be under his control.[8]

In other words, although Hungary was the first Central European state to have a civilian defense minister, no Hungarian civilian exercised effective oversight or control over Hungarian military matters (as was nominally exercised by civilian Deputy Defense Ministers Antonin Rasek since December 1989 in Czechoslovakia and Bronislaw Komorowski since April 1990 in Poland). The Hungarian defense ministry of 138 people[9] dealt mainly with social and political questions in line with matters that parliament was concerned. The commander of the Hungarian army and the armed forces remained separate and beyond Defense Minister Fur's purview. The Hungarians adopted a defense structure that was based upon the model of Finland. But Finland had a strong president and Hungary had a "weak" president, so the model did not quite fit.

These problems and political differences escalated into significant tension in civil-military relations, causing a constitutional dispute between the (AFD) president and (MDF-FKGP-KDNP) government and problems within the army's leadership. The constitutional debate involved questions over the sphere of authority between the commander of the Hungarian army and the defense minister, and ultimately between the president and the (MDF) prime minister. An October 1990 taxi drivers' blockade brought these different views about presidential and prime ministerial authority to a test. When Defense Minister Fur and Prime Minister Antall wanted to call up military equipment to remove cars from Budapest bridges, President Goncz, as commander-in-chief, refused and threatened a constitutional crisis. Though the prime minister and defense minister backed off, in an interview shortly after this incident Defense Minister Fur noted that the real issue was the relationship of the defense ministry to the armed forces:

> [O]ne of the important things to settle is the relationship between the [defense] ministry and the army commanders. The unclarified questions

emerge not so much in the relationship between the commander-in-chief, the ministry, and the Army, but rather in the relationship between the Army and the ministry.[10]

Soon after the blockade the government questioned the president's authority to command the army and initiated a review of the issue in the Constitutional Court.

During the spring of 1991, though, President Arpad Goncz (AFD), Prime Minister Antall, and Defense Minister Fur (MDF) still had differences of opinion over control of the armed forces. Lajos Fur argued that the leadership of the army was oversized, that it was unnecessary for the Hungarian Army Command and the General Staff to function in parallel, and therefore it would be desirable to adopt a leadership structure consistent with other European democracies.[11]

President Goncz countered in an interview that "attempts are being made to transform the Army by abolishing the command system, which I do not agree with . . . [adding that] the argument is not yet closed."[12] Tension reached such a pitch that LTG Kalman Lorincz, commander of the Hungarian army, submitted his resignation to President Goncz on 29 March 1991. Goncz, Antall, and Fur refused to accept Lorincz's resignation.[13] They recognized this civil-military issue to be a serious problem and mandated a new defense reform that was developed at the end of 1991.

Parliamentary Defense Committee member Bela Kiraly argued that the president was the clear commander-in-chief of the armed forces; yet the constitution placed two restrictions on his command. First, it authorizes the National Assembly to decide on deploying armed forces within Hungary or abroad. Second, it requires the prime minister's countersignature regarding every action involving national defense. Upon the National Assembly's declaration of war or emergency, presidential authority and responsibility expand. In sum, Bela Kiraly felt no constitutional change was required, but he argued that the commander of the Hungarian army position be abolished, that its responsibilities be transferred to the Hungarian CoS, and that the CoS be unconditionally subordinated to the defense minister.[14]

During 1991 two further tests brought the issue of military command to public attention—the failed Soviet coup and increasing problems along the Yugoslav border. The failed Soviet coup in August 1991 only partially tested Hungary's machinery, in part because the last Soviet troops from the SGSF had already left Hungary in June and the Warsaw Pact had disbanded on 1 July. (In contrast, Poland still had Soviet troops from the Northern Group of Forces on its soil.) When the National Security Cabinet met on 19 August to examine the situation, it noted that the borders were calm and concluded that Hungary was in no immediate danger. Prime Minister Antall met with members of the six legislative parties, who expressed full unity with the approach taken by the Cabinet that Hungary should take a restrained and moderate approach to the affair.[15] Hence, no military orders or special measures, which would have required a National Assembly vote, were ever issued.[16]

The second test involved the sporadic overflights of Yugoslav aircraft. Hungary's response also evidenced restraint. Despite the fact that no military mobilization measures had been issued and the heightened alert of Border Guard and Hungarian armed forces had been handled normally, on 18 September 1991 CoS LTG Janos Deak expressed concern to the National Assembly Defense Committee, arguing that if an emergency arose—for example if Hungarian barracks were attacked—LTG Kalman Lorincz lacked the authority to react rapidly. Deak argued that while Lorincz had mobilization authority, current constitutional stipulations presupposed that the decision either would be obstructed by the National Assembly (which requires a two-thirds vote) or would be made only very slowly.[17]

LAJOS FUR'S DEFENSE REFORM

Due to these external tests as well as increasing internal tensions between President Goncz and Prime Minister Antall, Defense Minister Fur in August 1991 sought an unequivocal Constitutional Court interpretation concerning peacetime direction of Hungarian Defense Forces. On 23 September, the Constitutional Court rendered its decision to limit presidential powers; it ruled that the president as commander-in-chief could *only* render guidelines to the military instead of issuing orders. The court concluded that the direction of the functioning of the armed forces was within the authority of the branch that exercised executive power (e.g., the prime minister and defense minister). It also ruled that the president is obligated to endorse candidates for state positions put forward by the government unless "the democratic operation of the institution in question is seriously threatened."[18]

In response to the Constitutional Court's decision, at the end of 1991 the defense ministry began a reorganization (see figure 3.1) to redress the problems created by the December 1989 defense reform. The new 1992 defense reform, which expanded the size of the defense ministry to over three hundred employees and accelerated personnel changes there, had the dual purposes of subordinating the military command to the defense ministry in accordance with the Constitutional Court decision and replacing career military officers with civilians in order to strengthen MDF control over the ministry. The new appointments increased civilian representation by reducing

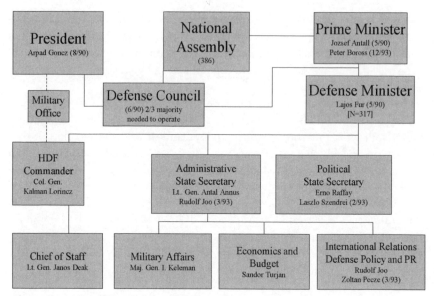

Figure 3.1.   Antall/Fur Defense Reform, 1992

the concentration of staff officers who had been Communist Party members in the defense ministry, replacing them in important midmanagement positions with civilians.

The President of the Republic Arpad Goncz remained the commander-in-chief of the armed forces with specific duties and responsibilities. Though the Constitutional Court defined the law, it remained untested in practice. HDF Commander Col. Gen. Lorincz remains subordinate to Goncz when the president is authorized to exercise his emergency powers during crisis and war. During peacetime Defense Minister Fur provided direction to Lorincz, who exercises command and control of the armed forces. Also subordinate to Fur is a political state secretary and an administrative state secretary, who supervises three deputy state secretaries.

By early December 1991 apart from Political State Secretary Erno Raffay and Deputy State Secretary Rudolf Joo, more new civilians were appointed to midmanagement positions in the defense ministry. Dr. Csaba Hende (MDF) became the ministerial parliamentary secretary, Dr. Zoltan Bansagi (MDF) headed the ministry's department for legal and administrative matters, and Laszlo Szoke (MDF) took over the department for social relations.[19] Thus, MDF personnel began to exert authority over the defense ministry.

Though the 1992 defense reform attempted to clarify the line of authority problems created by the 1989 defense reform, the issue of presidential versus prime ministerial authority during transition to war and during wartime remained untested. Different Hungarian views continued to exist as to whether the president would exert real (as against symbolic) powers during wartime. The debate was exacerbated by the untested role of the Defense Council, which was chaired by the president, but whose members also include the speaker and leaders of the political parties from the National Assembly and the prime minister, the ministers, as well as the commander of the Hungarian army and the chief of staff from the government.[20] Thus the powers of the president may be sharply curtailed by the predominance of political opponents on the Defense Council. Despite these nagging concerns, the 1992 defense reform had gone a long way to solve many peacetime problems that resulted from the 1989 defense reform.

In order to get further clarification on significant matters of disagreement, the eleven-member Constitutional Court became Hungary's locus of adjudication. Prime Minister Jozsef Antall asked the Constitutional Court on 25 May 1992 to rule on the president's scope of authority in firing government officials. The issue arose when President Goncz refused to countersign Antall's order to fire the director of Hungarian television, Elemer Hankiss. Antall asked the Constitutional Court to declare Goncz's obstructionism unconstitutional and, indeed, on 8 June 1992, the court, in a seven to three decision, ruled that the president could only block the prime minister's appointments and dismissals *if* legal procedures were not followed, the candidates were incompetent, or if in accepting the government's decisions Hungarian democracy would be threatened.[21]

In August 1992 a proposed draft national defense bill attempted to eliminate "management duplications" by expanding the government's management authority. Parliament (or, in case of declared emergency, the Defense Council) would be responsible for approving the basic principles of national defense, directions of military development, and the budget. During peacetime, all other decisions related to army mobilization, location, leading, and training of troops—as well as partial deployment in case of external threat and until parliament could decide—would come under government authority. The president would continue to have the title of commander-in-chief with no authority to command the armed forces. In peacetime his authority would be limited to approval of defense plans and to appoint and release high-ranking commanders in accord with the responsible minister's recommendation.[22]

In September 1992 the Constitutional Court approved Defense Minister Fur's request if he could fuse the Commander of the Hungarian Army position with the Chief of the General Staff without a constitutional amendment.[23] In February 1993, Fur submitted two related draft laws on defense to Parliament. One dealt with necessary constitutional changes; the other was the defense law itself.

The first law, which followed an April 1991 Constitutional Court decision[24] (adds Article 19[e] to the Hungarian Constitution), provides a new power for the executive. Under the new law, the government may, in three limited cases (invasion of Hungarian airspace, surprise air attack, or surprise invasion) order immediate military action of not more than two Army brigades (5,000 troops) without specific agreement from the president and without declaration of emergency by Parliament.[25] The government, however, is obliged to inform Parliament of any such decision.

The second law dealt with the organization of the Border Guard; defining the circumstances in which they fell within the jurisdiction of the military (as they did) or the police. The decision was necessary and significant because of the Yugoslav crisis. If the Border Guard were under the military, ultimate control would lie with Parliament; if they were under the police, then they would be under the executive, or minister of interior. In the bill, the Border Guard falls under the police, except for a state of war, and is subject to executive control. An overwhelming Parliamentary majority enacted both laws as amendments to the Constitution on 7 December 1993.[26]

In effect, Hungary had effectively negotiated the transition from Soviet Supreme High Command of its forces under the Warsaw Pact's Statute system to national command. Though initially confused by the December 1989 defense reform, effective Constitutional Court challenge and resulting defense reform clarified the chain of command problem.

Lajos Fur's Reform of the Armed Forces. By 1992, Hungarian Defense Forces (including civilians) comprised 100,000 from its 1989 size of 155,700 (see table 4.2). The number of eighteen-month conscripts declined from 91,900 in 1989 to 51,100; professionals from 30,500 to 22,900 (of which 8,500 were NCOs); and civilian employees from 33,300 to 26,000.[27] During 1992 Hungarian Defense Forces were reorganized; army brigades of a new type were created and the organizing and forming of mobile Armed Force units began.[28] In addition, a training center for peacekeeping forces was designed to train the first Hungarian peacekeeping company.[29]

The second phase in the Army's development would last until 1995, during which the forces were to be stabilized and conditions established for modernization after 1995, when funds were supposed to become available. The 1993 defense budget of Ft64 billion, which was increased to Ft66.5 billion in 1994, left very little room for modernization as 91.2 percent of the budget was needed for day-to-day operations.[30] Immediate aid for Hungarian Defense Forces came from Germany in April 1993, with its decision to supply spare parts, electronic equipment, and 20 L-39 training aircraft from former East German army stocks.[31] Arms also came from Russia, with the decision to supply 28 MiG-29s in October–November 1993 to cover $800 million of its $1.6 billion debt to Hungary.[32] The National Assembly earmarked Ft1.1 billion in 1993 to install 113 electronic Identify Friend or Foe (IFF) systems to the Hungarian Air Force that were to be installed on 28 MiG-29s, 9 MiG-23s, 65 MiG-21s, and 11 Su-22s during 1994.[33]

On 23 February 1993 Laszlo Szendrei (a Hungarian Democratic Forum MP) replaced Erno Raffay as political state secretary of defense.[34] On 31 March 1993 Rudolf Joo, an MDF civilian, replaced LTG Antal Annus as administrative state secretary, thus placing the defense ministry's top three posts in civilian hands.[35] On 14 April 1993 the National Assembly unanimously approved Resolution No. 27 concerning the Basic National Defense Principles of the Hungarian Republic.[36] Also on 7 December the National Assembly adopted (with 277 deputies for, 1 vote against, and 1 abstention) a new Defense Law to come into effect on 1 January 1994. According to the law, military service would be 12 months and civilian service in the military would be 18 months.[37]

In October 1993 when Boris Yeltsin survived a coup attempt in Moscow Lajos Fur noted that he survived in large part because "the Army, with its neutrality . . . unambiguously committed itself to support Yeltsin."[38] President Goncz noted that "I can promise one thing: I will never give the order to shoot on the Hungarian Parliament . . . [adding] the struggle in Russia will lead Hungary to work harder than ever for membership in the EU and NATO."[39]

In October 1993 the 88th Airborne Infantry Battalion was established as part of the Hungarian Defense Forces' restructuring. LTG Bela Gyuricza noted that its function was to make available to the military leadership a rapid deployment unit capable of preventing and managing armed conflicts and suitable to perform UN peacekeeping functions.[40]

On 12 December 1993 Prime Minister Antall died and the government's legal mandate ended. The interim government, under Interior Minister Peter Boross, initially operated with reduced powers until the president nominated a new prime minister, who had to be confirmed by a majority vote in Parliament. (Failure to appoint a government within forty days would result in new elections called by the president.) Parliament ultimately confirmed Boross by a majority vote.

On 14 January 1994, the Government announced that it would merge the defense ministry and the General Staff of the Army Command in accordance with the 7 December 1993 Defense Law, thereby placing the armed forces under civilian control in peacetime and war. This was scheduled to occur when General Kalman Lorincz reached the mandatory retirement age of 55 in February.[41] LTG Janos Deak, the CoS, assumed the post of Commander of the Hungarian Army on 1 March and was promoted to ColGen on 15 March. According to Lajos Fur, as of 1 March 1994, the defense ministry would have three state secretaries; political, administrative, and Chief of General Staff.[42]

In summary, the Jozsef Antall mandate focused on correcting many of the problems created during the Nemeth period. Constitutional amendments in June 1990 improved Parliamentary oversight of the armed forces and police by requiring a two-thirds majority for employing them. Though intelligence organs were restructured and brought under government control, the defense minister had concerns about the limits of his authority over the armed forces. Constitutional challenges led to a major defense reform in 1992 that brought the armed forces under the government's purview, correcting the problem created by the December 1989 defense law. Two crises during the period—the attempted Russian coup and Yugoslav over flights—contributed to other defense reforms in December 1993 to enhance the government's power to call up to 5,000 troops and then inform Parliament and to move the Border Guard to the Ministry of Interior. During the Lajos Fur mandate the defense budget spiraled downward from 2.5 percent of GDP to 1.7 percent. He implemented reorganization of the defense ministry, large reductions in the total armed forces from 155,700 to 100,000, reduction in the length of conscription service to 12 months, and acquisition of 28 MiG-29s.

4

1994 Parliamentary Elections: Post-Communist Return of Gyula Horn

On 30 June 1993, the Hungarian cabinet submitted a draft bill aimed at modifying the Electoral Law of 1989. It raised the electoral threshold from 4 percent to 5 percent and modified the procedure for by-elections. Now all by-elections would be held on the same day once every year, and never during the year of a general election.

With 69 and 55 percent of Hungary's eligible voters casting ballots on May 8 and 29, respectively, the 1994 parliamentary elections (like those in Poland in September 1993) brought the HSP back to power (see table 4.1) with an absolute majority; of the 386-seat Parliament, the HSP gained 209 (or 54 percent) of the seats for 33 percent of the vote. The second place AFD received 70 (or 18 percent) of the seats on a popular vote of 20 percent; followed by the former ruling MDF with 9.8 percent of the vote and 37 seats. Although the socialists had secured a parliamentary majority, they were concerned about legitimacy and decided for purposes of credibility to enter into negotiations and form a coalition government with the AFD. Thus, with 51 percent of the popular vote, the two HSP-AFD coalition parties had the necessary two-thirds parliamentary majority (279 seats or 72 percent) to amend the constitution.[1]

Hungary's 1994 disproportionate vote-to-seat ratio was remarkably similar to those in 1990, which resulted from Hungary's "mixed" electoral system. Out of the 386 deputies, 176 were chosen in two-round (majority and plurality), single-district elections, while up to 152 seats were filled in proportional votes in 20 regional constituencies, and at least 58 representatives were chosen from a national compensation list.[2]

Similar to Poland, one of the consequences of the Hungarian electoral system is that while the disproportion magnifies the strength of the winning parties and enhances their ability to govern, it is wholly unsuited when it

Table 4.1.   Hungarian Parliamentary Elections: 8 and 29 May 1994

Political Party	Seats	% of Votes
Government (N = 279)		
Hungarian Socialist Party (HSP)	209	33.0
Alliance of Free Democrats (AFD)	70	20.0
Opposition (N = 105)		
Hungarian Democratic Forum (MDF)	37	9.8
Independent Smallholders and Citizens Party (FKGP)	26	6.8
Christian Democratic People's Party (KDNP)	22	5.8
Alliance of Young Democrats (Fidesz)	20	5.3
Total	386	80.7

comes to the needs of constitutional politics. The constitution's amending formula, which allows two-thirds of Parliament to revise the constitution, should not be left as it is. Theoretically, the new socialist-liberal (HSP-AFD) coalition could unilaterally act under the inherited amending formula to change the constitution along with the current two-thirds electoral law and permanently undermine the chances of the weak opposition.

FAILED CONSTITUTION EFFORTS

Both the electoral law and the constitution's amending formula presented dangers to Hungary's parliamentary democracy and constitutional stability. One indication of this danger occurred on 30 September 1994 when all four opposition parties walked out of Parliament as voting began on a constitutional amendment to voting procedures for local elections. But the HSP-AFD coalition, with two-thirds majority, voted to change the constitution to simplify procedures for local election and improve the chances of the incumbent left-of-center majority. This led to opposition charges of a constitutional dictatorship.[3]

The procedures for amending the constitution needed to be changed to bring the past five-year transition period to a legal close. Though Hungary needed a new procedure to prevent ceaseless parliamentary tinkering with the constitution, the new socialist-liberal coalition was *not* in a good position to initiate a new phase of constitution making. First, although it held 72 percent of the seats, its 51 percent electoral base was too narrow to establish anything but a winner's constitution. Second, neither the HSP (whose forerunner HSWP imposed a pseudo-constitution on the country in 1949) nor the AFD was well situated to sponsor a new constitution. But Hungary needed to revise its constitution to deal with clarifying the role of the president, reducing the Constitutional Court's powers, and redefining the role of the prosecutor.

During 1995, Hungary planned to draft and adopt a new Hungarian constitution because the coalition parties were committed to this goal. They formed a twenty-seven-member parliamentary committee (HSP would have ten; AFD five, and opposition parties ten) to draft the new document, working under the minister of justice. Though they planned to present the new constitution for popular ratification by 20 August 1995 (when new presidential elections were required),[4] the process took longer than expected and was postponed beyond the time frame of the government's mandate.

Several items on the constitutional agenda included presidential powers, guaranteeing judicial independence by a National Judiciary Council, redefining the role of the public prosecutor, reforming local government, and trimming the Constitutional Court's functions.[5] A new constitutional amending formula was proposed, requiring a second parliamentary session to ratify amendments made by a previous one. Finally, a new electoral law would be proposed, abolishing the second electoral round, keeping a mixed system, but taking the principle of proportionality into account.

During the fall of 1995 the six-party parliamentary committee had been at work preparing the new constitution. Though the Justice Ministry's draft did not foresee any major changes to the existing constitution, they needed to develop more precise language for the powers of the president and decide whether the prosecutor's office should be independent or under the government, and whether Parliament should remain unicameral or become bicameral.

Because the HSP-AFD Socialist-liberal coalition had 72 percent of the seats and only two-thirds vote was necessary for ratification in order to achieve a broad-based consensus, the six parliamentary parties agreed to submit the draft to Parliament *only* if at least five of the six parties could agree. Though the intention was to submit the draft constitution to Parliament before the end of 1995 so as to be in place in 1996, delays occurred, pushing this to the end of the government's mandate.

Despite the good intentions, consensus proved elusive. On 6 June 1996 Parliament ended debate on the draft constitution. While the ruling coalition wanted the final text by December 1996, the opposition parties now just wanted amendments to the present constitution and to postpone the draft until after the 1998 elections.[6] When Parliament began debating the new draft constitution on 19 June 1996 the greatest divisions involved the inclusion of social rights provisions, with two parties—the AFD and Young Democrats—opposed.[7]

In sum, an objective that had initially appeared to have a clear agenda and a six-party national consensus turned into a much longer and more difficult process. Developing a broad-based consensus constitution that was not perceived as a victors' *diktat* proved to be more elusive than all sides had originally envisioned. As a result, the new constitution became a significant item of unfinished business for the Horn government.

During 1997 concerns about the failure to adopt the new constitution and continually amend the document were often raised. Former AFD Deputy Mikos Haraszti noted that if the government majority (72 percent) keeps amending the constitution without the opposition's support—like Vladimir Meciar in Slovakia and Alexander Lukashenko in Belarus—consensus eventually could break down. In the end, "continual tinkering implies the possibility of a Constitutional dictatorship."[8]

In January 1998 the government did present Parliament with a bill amending the law on the Constitutional Court. The bill proposed to elect judges for twelve years (rather than nine), and to extend the mandate of the existing court by three years. The possibility of reelecting judges would be eliminated, and the age limit would be seventy years.[9] Of particular note, a 1998 Hungarian survey indicated that the Constitutional Court enjoyed the most confidence (exceeding 60 percent), while the government had the least confidence with only 35 percent, and the Hungarian Defense Forces had 50 percent support.[10]

The National Assembly, through its nineteen-member Defense Committee, was slowly attempting to develop oversight of the military through budget, approval of the Basic Principles of National Defense and Defense Bill, and deployment of armed forces. The Defense Committee had opposition representation (with six members) and continuity (with seven members remaining from before the 1994 elections), to include former Defense Minister Lajos Fur and two retired generals. Also, parliamentary MPs could not be members of the military, thus ensuring civilian control.[11]

The Parliament Defense Committee, though, had limited effectiveness and lacked staff support. Its limitations were most apparent in oversight of the defense budget as manifest by Defense Minister Keleti, who made unilateral decisions to buy T-72 tanks from Belarus with funds derived from the sale of defense real estate and sent MiG-29s to a PFP exercise in Poland without proper consultation. In other words, Keleti's actions denied Parliament the opportunity to deliberate as to whether buying T-72s was the best way to expend public funds or if those funds could be better directed to other priorities, such as readiness and training.

Parliamentary oversight began to improve slowly. For example, preceding the 30 June termination of the Stabilization Force in Bosnia (SFOR), Parliament on 24 February 1998 decided to approve Hungary's SFOR commitment.[12] In addition, after the NATO Madrid decision in July 1997 the Defense and Foreign Affairs Committees of the three NATO invitees decided to meet periodically and discuss cooperation on military and security matters, crisis management, and defense infrastructure and acquisitions. They had agreed to closely cooperate on their EU integration as well.[13]

GYORGY KELETI'S DEFENSE REFORM: "FROM CITIZENS IN UNIFORM TO GENERALS IN SUITS"

When the new government was formed after the 1994 election, HSP leader Gyula Horn became prime minister (see figure 4.1). On 24 June the HSP-AFD coalition signed a government agreement; the AFD would take over three ministries—interior, transportation, and education—and the HSP would take over the remaining twelve. Gyula Horn appointed retired Colonel Gyorgy Keleti as new defense minister. Keleti, former press spokesman for the ministry under Lajos Fur, had left under a cloud in 1992. Keleti noted that he walked out on former Defense Minister Fur "because the conditions prevailing in the ministry made it impossible to work normally with the minister and several of his employees."[14] After leaving the military, Keleti had been elected to parliament twice from an individual electoral district in a 1993 by-election and then again in 1994.

Upon Defense Minister Keleti's return on 15 July 1994, he began to replace all the MDF personnel mostly with former colleagues from the

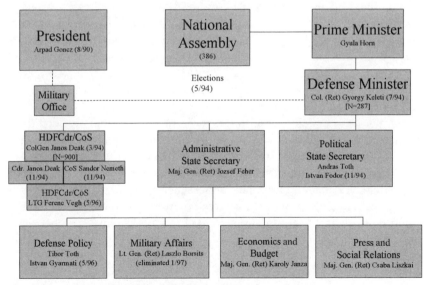

Figure 4.1. Horn/Keleti Defense Reform, 1994

Hungarian Defense Forces; Col. (reserve) Jozsef Feher was promoted to major general on 15 July and appointed administrative state secretary;[15] retired LTG and former CoS Laszlo Borsits and MG Karoly Janza became deputy state secretaries. Keleti also appointed MG Csaba Liszkai to supervise press and social relations with the rank of deputy state secretary. The military also took over departmental-level positions. Col. Peter Haber, an old colleague of Borsits, became head of the Military Department under Borsits; Col. Nandor Gruber replaced civilian economist Sandor Kovacs as head of Defense Economic Department; and Col. Istvan Szekeres replaced civilian sociologist Laszlo Dobos as head of Department on Social Relations and Culture.[16]

One civilian, Andras Toth became political state secretary; but in November, Dr. Istvan Fodor replaced Andras Toth (who moved to head the prime minister's office) as the new political state secretary at the defense ministry. The only other high-ranking civilian was Tibor Toth (an expert on disarmament from the foreign ministry) as one of the three deputy state secretaries. Another civilian expert Istvan Gyarmati replaced Toth in May 1996.[17]

Keleti also began an internal reorganization of the defense ministry, cutting it from 317 to 287 people, noting that the structural reorganization and reduction of the defense ministry would result in an increase in the percentage of civilian employees from 35 to almost 40 percent.[18] Defense ministry spokesman Col. Lajos Erdelyi noted that the reorganization was "an internal affair," adding that according to law, the defense minister can make such decisions. In response, Imre Mecs (AFD), chairman of the National Assembly's Defense Committee, expressed concern about "militarization" of the defense ministry and noted that there was not enough "civilian staff."[19]

In an early interview Defense Minister Keleti noted that he was sure that he would have harmonious relations with the generals and that he intended to act as a civil politician and not a "former colonel." He also indicated that he intended to abolish the government order that provided deadlines for the organizational fusion of the army headquarters with the defense ministry, and after further study, decide whether or not the proposed fusion was really justified, since "the Army leadership should receive sufficient independence to plan and lead their professional activity."[20] Keleti added that he met with President Goncz and agreed to meet with him once a month to inform him about the army's situation, and that he also invited Robert Pick, head of the president's military office, to attend all cabinet sessions of the ministry.[21]

Keleti then reorganized the General Staff, providing it with more authority in military planning, including intelligence. In early September 1994,

Defense Minister Keleti made his decision and recommended to the National Assembly Defense Committee *not* to merge the defense ministry and army headquarters.[22] As of 1 November 1994 Keleti once again divided the two top army positions when he appointed LTG Sandor Nemeth to become CoS, while retaining Janos Deak as commander of the Hungarian Defense Forces.[23] Keleti wanted the division of the post to "contain" the commander from "above" and "below."

That Defense Minister Keleti's decision to separate the General Staff from the defense ministry needed to be revisited was stressed by a British study team review of the Hungarian Ministry of Defense in February 1996,[24] and then also by Keleti in April when he noted that "we must put integration on the agenda sooner or later."[25] Keleti suggested that the first step might be to leave unfilled the position of commander of the Hungarian army and eliminate the function, then create a new headquarters later in 1996, and then perhaps by 2000 make the headquarters part of the defense ministry.

Defense Minister Keleti noted that his major concern was retention of professionals in the Hungarian Defense Forces. Because the army could not be financed from the budget with its current structure, Keleti proposed to reduce personnel by retaining professional officers and to call fewer conscripts to service; some two thousand less in August 1994, with repeated reductions in February 1995 and 1996 reducing the army by twelve thousand overall.[26] In addition, Keleti promised that he would continue to reduce the intake of conscripts and beginning in 1996 would reduce their national service time to nine months[27] with a more intensive training program.[28]

Keleti noted that ministry calculations indicated that it needed Ft69 billion in 1994, of which 7.2 billion was not covered by revenue; and that efforts by defense managers could only cut 3.5 billion by cost-saving means.[29] Admitting to deficiencies in the ministry's budgeting process, MG Karoly Janza, deputy state secretary for Economic and Budgetary Affairs, noted that the financial situation was worse than he expected, citing significant infrastructure expenses of more than Ft1 billion to maintain the recently acquired MiG-29s. Janza suggested that reducing exercises and conscripts was the only way to reduce the shortfall.[30]

When Defense Minister Keleti announced in September that he would cut the size of the defense ministry as a cost-saving measure, he noted that he would retain the Army Command size (of roughly nine hundred) because intermediate command levels were to be eliminated and the military zones were to report directly to the General Staff.[31] The reform would reduce Hungary's four military districts to two, resulting in a further reduction of staff.[32] In support of Keleti, retired Colonel Peter Deak added that the military needed fewer levels of command to enhance the military's ability to react rapidly, make communication more rapid, and create better informed and more independent staffs. Deak, though, also noted that the rear services organizations were not functioning properly and that the General Staff was too big and appeared to duplicate the defense ministry.[33]

On arms acquisitions, Keleti also shifted further from his predecessor's policy. After the MiG-29 acquisition from Russia, Defense Minister Lajos Fur had indicated that he also would like to get the S-300 missile air defense system in exchange for some of the remaining $800 million debt. Defense Minister Keleti rejected this policy; instead, he wanted to acquire BTR-80 armored personnel carriers for the Hungarian army and Border Police and spares for the 28 MiG-29s for $320 million of Russia's $800 million debt.[34] But Russian deliveries were postponed from November 1995 to the spring of 1996. In the end, the Hungarian army was to receive 450 BTR-80s (of which sixty-eight would be used by the Border Police), 400 METIS armor-piercing missiles and 15 mobile launchers, and $50 million worth of MiG-29 engines and spares.[35]

Keleti also succeeded in his efforts to acquire military spare parts from Ukraine in return for Hungarian medicines.[36] Finally in March 1996, despite opposition from some members of the Parliament's Defense Committee, because Defense Minister Keleti had failed to consult with them beforehand and because of increasing dependence on Russian arms manufacturers, the Hungarian army decided to purchase one hundred T-72 tanks from Belarus at a very favorable price.[37] Presumably, the T-72s would be cheaper than modernizing Hungary's T-55s, and the Defense Ministry proposed paying for the T-72s with money obtained from selling real estate.[38]

Defense Minister Keleti also stressed that he wanted to modernize the air force, particularly ground-based radars.[39] Also following on Fur's success to acquire twenty L-39 trainers in 1993, Keleti successfully acquired twenty Mi-24 combat helicopters from Germany, which were inherited from the former East German National People's Army.[40] Although CFE limits Hungary to 108 helicopters, after the German installment, Hungary's inventory increased to only 59, since some of the German supply were dismantled and used for spare parts.[41]

ARMY REFORM AND SHRINKING DEFENSE FORCES

When Army Commander Janos Deak presented the army reform concept to the National Assembly in January 1995, he noted that the program was motivated by the fact that budgetary allocations were inadequate to maintain existing military structures and by the need to modify the Hungarian armed forces in order to integrate into NATO.[42] In sum, the reform was being pulled in two directions.

On 1 March 1995 Prime Minister Gyula Horn announced the new organization of the top leadership of the armed forces (see figure 4.2). Subordinate to the commander of the Hungarian army Col. Gen. Janos Deak was the General Staff headed by LTG Sandor Nemeth and four Division Commands (MG Jozef Wekerle of Land Forces, MG Tibor Szegedi of Air Force and Air Defense, MG Nandor Hollosi of Logistics, and MG Janos Gilicz of Human Resources).[43] On 22 March CoS LTG Sandor Nemeth told the Parliament Defense Committee that the armed forces were preparing to establish rapid deployment battalions, comprised mainly of professionals and soldiers under contract, which later would be developed into brigades.[44]

In accordance with National Assembly Resolution 88/1995 on the Direction of the Long and Medium Term Reform of the Hungarian Defense and Its Personnel Strength,[45] the defense ministry issued a directive on 15 July 1995. In mid-September Deputy State Secretary Laszlo Borsits briefed the National Assembly Defense Committee that the Command of the Land Forces and the Regional Military Commands would be abolished and replaced by a Mechanized Army Corps Command. In addition an Air Force and Air Defense Corps Command would be set up and a Central Organizations Command would replace Budapest Regional Military Command by the end of 1995.[46]

One result of the military reorganization was the need to form brigades that could be deployed quickly; the first to be created was the 25th Mechanized Rifle Brigade in Tata. One effect of the reorganization was that between 1996–1998, the defense ministry's 19,200 civilian personnel had to be *reduced* by 69 percent to 7,800, the 14,400 professional military officers

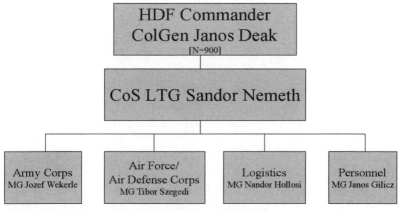

Figure 4.2.   Hungarian Commander/General Staff, 1995

by 37.5 percent to 9,000, and the 37,595 conscripts by 13 percent to 32,400. The number of NCOs would *increase* from 9,700 to 10,800. This demobilization would cost Ft3.2 billion in 1996, Ft1.3 billion in 1997, and Ft1.4 billion in 1998.[47] In sum, the four-year Keleti era would be distinguished by significant demobilization of the civilian defense employees and professional officer corps.

Financial constraints, though, continued to be particularly severe. Defense Minister Keleti noted on 8 March 1995 that during the previous four years the Hungarian armed forces had consumed their reserves of fuel, spare parts, and clothing, and that nothing remained, thus, replenishing the reserves had become an urgent priority. As a result, the armed forces had abandoned military exercises above the company level, the reserves were not called up in the second half of 1994, and air defense missile and air combat exercises had been "scrapped."[48] To save money airmen flew only forty-five to fifty hours per year.[49] This resulted in eighteen military aircraft accidents in 1994, and four accidents through September 1995.[50]

Col. Gen. Janos Deak admitted in a November 1995 report to the Parliament Defense Committee that the effects of reorganization, downsizing, smaller budgets, and reduced training were undermining morale. Deak noted that the situation with the professional staff was "particularly critical . . . [that] the officers' salary is not in proportion with their responsibility. The army does not have money for technical development, nor even for gasoline sometimes."[51] This problem had still not been resolved by April 1996. After the Parliament's Defense and Security Committee heard from defense ministry and Hungarian army leaders that seven hundred professional soldiers—mostly experts—had left the army since the summer, it concluded that the armed forces' reform needed modification because the present ideas could lead to "crisis."[52]

The 1995 defense budget of Ft77 billion (1.5 percent of GDP) was further burdened by a Ft2.4 billion carryover from 1994. According to Defense Minister Keleti, the "real value of the Army's budget has decreased by 58 percent since 1990."[53] In order to deal with the tight budget, the defense ministry had to lay off three thousand civilian employees as of July 1995 and a further three thousand over the next three years.[54] Deputy State Secretary for Defense, LTG Karoly Janza noted that in 1996 the defense ministry would need Ft80 billion just to function and a further 10–15 billion to launch the reform process.[55]

Fiscal restraints contributed to very different views about the size of Hungarian armed forces. In 1989 Hungary had started with total armed forces (military *and* civilian) of 155,700. In January 1996 the total size of the armed forces was 80,902 military and civilian personnel; with an active military component of 61,694 (including 37,595 conscripts).[56] By the beginning of 1997 the total size was to decline further to 69,812, requiring dismissal of 5,000 civilian employees, 4,000 enlisted personnel, and 3,000 officers.[57] According to Peter Haber, head of the Military Department of the defense ministry, the long-term reform concept envisioned a total active force of 60,000 by 1998; an army of 52,200 soldiers (including 32,400 conscripts) and 7,800 civilians. The army would thus comprise 0.5 percent of Hungary's population (see table 4.2).[58]

But these plans would be revised. In May 1995 retired General Janos Sebok and member of Parliament had a very different view noting that the "current size of the Hungarian Army could safely be reduced by at least 50 percent . . . [adding] I advocate a reduction of the Army's size to 40,000 men and that the reduction should be implemented by 1997."[59] By May 1996 fiscal constraints appeared likely to continue to drive manpower levels lower. The Hungarian government's state administration reform plan required cutting government expenses through 1998 and proposed reducing the army's peacetime personnel size to thirty thousand.[60] In sum, while it remained ambiguous just how small Hungary's defense forces would actually become, it was certain that they would become smaller. Also it was

Table 4.2. Hungarian Armed Forces as of 1 January

Year	Total	Conscripts (mo.)*	Career	Contract	Civilians
1989	122,400	91,900 (18)*	30,500		33,300
1990	110,700	81,000	29,700		32,500
1991	94,000	65,300	28,700		27,600
1992	74,000	51,100	22,900		26,000
1993	74,340	52,340	22,000		25,660
1994	73,660	51,560 (12)*	22,100		24,060
1995	68,261	46,350	21,911		23,894
1996	61,694	37,595	24,099		19,208
1997	55,731	34,385 (9)*	21,346	2,200	14,081
1998	53,700	33,000	20,700	4,000	11,100
1999	52,200	32,400	19,800	4,300	7,800
2000	38,000	13,000	19,900	5,100	7,500
2001	36,400	13,000	19,900	3,500	7,500
2002	36,170	13,260 (6)*	18,910	4,000#	6,100

= estimate
* = length of conscript term in months
Sources: National Defence '95 and *National Defence '96* (Budapest: Ministry of Defense), 13, 27, respectively; *Reform of Armed Forces: 1995–1998–2005* (Budapest: Ministry of Defense), 11. Also for an excellent study that conforms with the above data, see Laszlo Zoltan Kiss, "The Hungarian Army: Professional Soldiers and Conscripts," in Tamas Kolosi, Istvan Gyorgy Toth, and Gyorgy Vukovich (eds.), *Social Report* (Budapest: Tarki Social Research Informatics Center, 1999), 189–91. Most recent figures provided by Arpad Szurgyi in Budapest, April 2002.

clear that their ultimate size was not being determined by a mid- to long-term strategic plan, but by scarce resources.

On 27 July 1995 Jozsef Feher expressed optimism with progress on the bill on the Legal Status of Armed Forces professional personnel. He indicated that the long-awaited bill would be put before Parliament in October and could become a legally binding statute in January 1996. The bill would reconcile service status with the legal system and would provide a framework for interest representation within the armed forces, resolve anomalies of qualifications and promotions, clarify the rights of professional soldiers, and establish salaries and retirement benefits in line with those of other public employees.[61]

On 5 October 1995 the government discussed the bill on professional soldiers and began coordination talks. However, when the bill got bogged down, professional soldiers collected signatures for a petition to Parliament protesting the continued delay of the law to settle their legal status.[62] When the law was finally passed in early May 1996, four thousand professional soldiers in the Soldiers' Trade Union expressed dissatisfaction claiming that it excluded them from public life, that they could not even take on deputy posts in the local government, and that the government kept delaying the deadline for introducing a new army wage table and pay raise.[63] Even though the salaries of soldiers had been raised 19.5 percent in January 1996, many soldiers continued to leave the army.[64]

Jeno Poda (MDF) of the Parliament's Defense and Security Committee described the "personnel and technical conditions of the Hungarian Army as tragic,"[65] criticizing the fact that over one-half of the defense budget was spent in Budapest and that most reductions in personnel had occurred in the provinces and in combat units.

PARLIAMENTARY/DEFENSE
MINISTRY CIVILIAN OVERSIGHT

Civilian control of the Hungarian army was again raised as an issue in the spring of 1995. Criticizing increased militarization and "deficiencies" in the defense ministry, Parliamentary Defense Committee Chairman Imre Mecs (AFD) noted: "The executive should control military matters, but this is not done with the necessary effectiveness, so the National Assembly's Defense Committee has to reinforce its supervision in this domain."[66] Keleti responded by noting that parliamentary control over the army is already strong, but admitted "we have not really managed to establish the defense ministry's civilian basis in the past four years. At the moment, 40 percent of the ministry's 287 employees are civilians."[67]

Administrative State Secretary of Defense Jozsef Feher, noting that though there had been much progress made in establishing civilian control since the change of regime in 1990, admitted:

> However, the administrative framework of the Army which makes it possible for civilian observers, legislators, state administration employees, and society to see clearly how the Army uses the money entrusted to it and, what is more important, how it uses the people has *not* been created.[68]

Feher added that the army still needed to establish a budget planning system, a military defense planning system, and legal regulations.

Tamas Wachsler (Young Democrat-Fidesz) of the Defense Committee expressed a lack of confidence and also complained that the defense ministry is often unwilling to provide secret information to the members of the committee, while Imre Mecs argued that the authority of the Defense Committee needed to be expanded.[69] The obstacles to the proper functioning of the Defense Committee lie in the lack of expertise of new members and in the absence of professional support staff or advisors on military matters. This leaves great space for maneuvering of the military, which sometimes lacks good will toward parliament. Wachsler noted "unless MPs ask *the* correct question, they will not get the answer they are looking for."[70] As a result, they are hardly able to initiate legislation; usually the Defense Committee follows the army's and defense ministry's initiatives.

When Defense Minister Keleti discussed draft legislation on the rights and obligations of conscripts and announced in April 1996 that at some time in the future conscription would be reduced from one year to nine months, Imre Mecs of the Defense and Security Committee announced that this would *not* occur because the country's defense capability needed to be considered and that such a move would require more money.[71]

Following on the heels of the controversial T-72 acquisition from Belarus when Defense Minister Keleti failed to ask Parliament for prior approval, during May 1996 another major issue arose regarding the parliament's lack of oversight in the deployment of eight MiG-29s to a NATO-PFP military exercise in Poland. Apparently the Aviation and Air Defense Command and Defense Ministry Deputy State Secretary Borsits failed to check and meet the constitutional conditions for deploying forces abroad. Hence, the state secretaries and defense minister were not informed. Amid parliamentary opposition leaders' cries for Defense Minister Keleti's dismissal, Prime Minister Gyula Horn defended his defense minister claiming that Keleti shared no personal responsibility in the matter.[72]

At the end of June Defense and Security Committee Chairman Imre Mecs complained that the "defense ministry was not fully aware of the Parliament's importance [adding that] the purchase of T-72s should have been coordinated with the Committee in advance, and not presented with a fait accompli."[73] Some of these differences were worked out in early July with parliamentary approval of Hungarian troop participation in exercises through 1996.[74]

Although the Defense and Security Committee's oversight of the defense budget still remained limited, it was one of the more effective Central European parliamentary defense committees, comprising five subcommittees with varying degrees of activity and effectiveness. First, the Special Investigations Subcommittee investigated secret service operations, the MiG-29 deployment to Poland, defense ministry treaties with foreign governments, and contract tenders. Second, the Budget Subcommittee held hearings on contract tenders for radars, short-range missiles, fighters, and IFOR procurement. Third, a Supervisory Subcommittee oversaw legal implementation of economic and social issues. Fourth, a Foreign Affairs Subcommittee that remained somewhat inactive. Fifth, an Ombudsman Subcommittee investigated complaints of internal violations.[75]

In belated response to some of these parliamentary concerns about the defense ministry, effective 2 January 1997, Defense Minister Keleti dismissed Laszlo Borsits as deputy secretary of state, eliminated the military affairs position, and transferred its responsibilities to Administrative State Secretary Jozsef Feher.[76] The defense ministry budgeting function needed improvement, particularly for NATO. Since the Hungarian state budget law did not require military expenditures to be linked to specific tasks, the 1997 defense budget of Ft96.8 billion was only four pages in length. According to Zoltan Martinusz of the defense ministry NATO Integration Department, NATO defense planning required Hungary to provide more information. Hence, the 1998 defense budget would have to be substantially more detailed and transparent.[77] In sum, defense ministry functions of multiyear defense planning, rational acquisitions and manpower development, transparent budgeting, and legislative liaison were *not* well developed.

KELETI'S "TWO CHANNEL" ARMED
FORCES MODERNIZATION

On 19 September 1995 the National Assembly unanimously approved two proposals on air defense to improve radars in two phases (phase one to the year 1998 would cost Ft20 billion) and low altitude air defense missile systems in two phases (fifteen air defense units to 2000 would cost Ft10–11 billion; thirty air defense units to 2005 would cost Ft20–21 billion).[78] According to Defense Minister Keleti the 1995 staff reduction made it possible for the army to implement a Ft10 billion development program during 1996 compared to the Ft2 billion spent in 1995.[79]

The United States agreed to provide $6.25 million in the modernization of radars in the framework of PFP.[80] When Westinghouse won the Ft10 billion tender for the radars, there were allegations of corruption. Parliament's Defense and Security Committee investigated the charges, but made little progress in substantiating them.[81] The new air defense system would reduce energy and staff requirements by a few thousand soldiers.[82] Since it cost about Ft300 million a year to maintain one thousand soldiers, the five thousand soldiers serving in the air defense radar system cost Ft1.5 billion annually. According to Col. Gen. Janos Deak, since the modern radar system would reduce manpower requirements by 25–33 percent (1,250–1,650 soldiers), the Ft1 billion annual savings would help amortize the costs of the system.[83]

Aircraft acquisitions would also have an impact upon the defense budget and force structure. While CFE allowed Hungary to have 180 combat aircraft, Hungary planned to keep only seventy to ninety by the end of the decade. Decisions were to be made in December 1996 for the purchase of thirty combat aircraft from among Swedish JAS-39 Gripens, US F-16s or F-18s, or French Mirage-2000s.[84] The issue became further complicated when the Israeli Aircraft Industries in May 1996 offered to refurbish twenty-eight of Hungary's aging MiG-21s for less than 10 percent of the cost of buying new aircraft.[85] In the end, when the finance ministry concluded that no deal could be concluded because no money was available, the government decided to postpone the decision until mid-1997.[86]

Defense Minister Keleti noted that Hungary's modernization was on "two channels." The air force and air defense modernization would likely be Western to better fulfill NATO compatibility requirements. Modernization of land forces, though, was another matter. Since there was no NATO standard for tanks and APCs, only for the built-in electronics and telecommunications systems, these could come from the East. Hence Keleti justified the acquisition of one hundred T-72s (for 5 percent of the cost of a new tank) from Belarus as *enhancing* Hungarian independence, since T-72 spares were produced in Slovakia, Ukraine, Belarus, Poland, and Russia.[87]

FERENC VEGH: EFFECTIVE DEFENSE
FORCES OR "OPERETTA ARMY"?

Toward the end of 1995, Defense Minister Keleti noted that the ministry was preparing comprehensive leadership changes in the coming year. On 1

December 1995 among the eleven generals who retired at the mandatory age of fifty-five years were Administrative State Secretary MG Jozsef Feher, Deputy State Secretaries LTG Karoly Janza, and MG Csaba Liszkai; however all three would remain in their present positions, but as civilian employees in the defense ministry.[88]

During 1996 another ten generals would retire because of mandatory age limits. Among them were Army Commander Col. Gen. Deak and CoS LTG Nemeth. In accordance with the parliament's decision of reorganizing the Hungarian army, the two positions would again be unified as they had been under Defense Minister Fur.[89] LTG Ferenc Vegh, who had been trained at the U.S. Army War College, became first deputy CoS in December 1995 and assumed command of the new fused position of army commander/CoS in May 1996.[90] In addition, MG Lajos Fodor, who had been trained at the U.S. National War College, became First Deputy CoS on 1 July 1996.

General Vegh would be responsible for restructuring and reorganizing the General Staff with the goal of developing a simpler and more efficient structure and reform of the military. The staff would be reorganized with the involvement of civilian experts, scholars, and NATO specialists.[91] In his testimony to the Parliament Defense Committee, Vegh noted that an army of 45,000 would be enough to defend the country.[92] He also openly noted that the army staff and leadership of troops was "hamstrung by customs and centralization [which he characterized as] dangerous." Vegh expressed impatience in that he could "not see clearly whether the generals really want changes to happen" and concern that among the troops "pessimism reigns supreme."[93]

Ferenc Vegh opened his tenure on 6 June noting that the Hungarian army had no more reserves and could no longer meet its obligations. Vegh noted that the annual cost of the army structure adopted by Parliamentary Resolution 88/1995 should be Ft143.5 billion; and that the army was in a very bad condition because it had only been allotted one-half that amount. Therefore, he challenged the politicians to decide whether they wanted to have an "effective defense force or an operetta army."[94]

Ten days later Vegh made clear the challenge to the politicians. The CoS noted:

> We will face unpaid bills this year. A political decision is needed to determine the types of tasks the Army should be capable of fulfilling and the budget needed for this. If we have this, we soldiers will be able to explain what kind of Army we are capable of activating under the given circumstances.[95]

In apparent agreement, Defense Ministry Deputy State Secretary Istvan Gyarmati admitted that: "Maintaining an Army of 60,000–70,000 will require about twice as much money as the current budget. For this reason it does not appear realistic."[96]

NATO: PFP AND IFOR/SFOR PEACEKEEPING

At the end of March 1995, Keleti noted that training of the first Hungarian peacekeeping force had been completed and was expected to deploy to Cyprus in the second half of 1995.[97] In the fall thirty-nine Hungarian troops did join an Austrian peacekeeping unit in Cyprus under the United Nations, and by the spring of 1996, the Hungarians increased their contribution to the staff to 120.[98]

Based upon the PFP exercise experience and military exercises with NATO states, Parliamentary Defense Committee chairman Imre Mecs noted:

> [W]e have a long way to go to catch up in the fields of telecommunication, organization, and cooperation, including the knowledge of languages. [Nevertheless, he concluded that] the Hungarian Army would be suitable for NATO membership around 1998.[99]

During 16–20 October 1995 a German-British-Hungarian PFP exercise (code-named Cooperative Light) was held. The exercise cost Ft150 million, with Hungary putting up Ft90 million, included 1,200 Hungarian and 350 foreign troops, and provided Hungary with the first opportunity to participate in the planning of a NATO exercise. MG Lajos Urban noted the difficulty of preparing the Hungarian staff commanders for six months for the exercise.[100] Also during 22–24 July 1996 a fifteen-nation search and rescue PFP exercise (Cooperative Chance-96) comprising 540 foreign and 600 Hungarian soldiers, took place in Hungary. Hungary covered Ft100 million of the exercise costs, the remainder covered by participants.[101]

When the conflict flared in the Balkans with the Croatian offensive launched on 4 August 1995, the Hungarian government strengthened border defenses and heightened the air force's readiness, to include placing their MiG-29s in readiness.[102] After the Dayton agreement was concluded in autumn, Hungary agreed to permit the United States to set up an army service corps of one to two thousand troops from a logistics unit in the vicinity of Pecs in southern Hungary to service the Bosnia Implementation Force (IFOR) for one year.[103]

The United States would pay rent for the army establishments and their refurbishment at Kaposvar and Taszar, as well as the costs of making the Taszar Airbase NATO-compatible.[104] Parliament approved this move on 28 November 1995. Defense Minister Keleti openly expressed his hope that the facilities would remain in Hungary after the IFOR mission ended as an "advance" for the time of "even more intensive contacts with NATO."[105] The Taszar Air Base, used for the IFOR and successor SFOR and supported by 3,500 troops, would be suitable for Hungarian military purposes, NATO standby tasks, and civilian air traffic.

In addition, on 27 March 1996 the United States signed an agreement, in line with NATO's Status of Forces Agreement (SOFA) program, that would cover 75 percent of the costs of damage caused while serving in Hungary, while the Hungarian Defense Ministry would sustain 25 percent.[106] Cooperative relations existed among the American and Hungarian troops at Kaposvar and Taszar,[107] though some problems appeared among civilians. Hungarian employees formed a trade union and complained of unpredictable work conditions and low wages,[108] as well as of sexual harassment.[109]

On 5 December 1995 Parliament passed a separate decision to send five hundred bridge-building troops to Bosnia for one year; though with some political opposition from the FKGP. Finances estimated to cost Ft1.5–2 billion ultimately limited the Hungarian contingent to four hundred troops.[110] The troops left Hungary for Okucani, Croatia, in two groups on 30–31 January 1996. In addition, thirty-five Hungarian policemen went to Bosnia as part of an unarmed UN peacekeeping mission.[111] In IFOR's Stabilization Force follow-on, a 310-man engineering battalion participated in Okucani. As the operation extended in time, rotation of Hungarian troops became necessary in December 1997, putting stress on the Hungarian army.[112]

Though SFOR participation was a financial burden for the Hungarians, they also derived benefits. According to General Tibor Cserepes, the NATO-led Stabilization Force paid the Hungarian army $60 million in 1997 to use the military base. During 1996–1997 the American troops stationed at Taszar spent $115 million.[113] (The entire Hungarian defense budget in 1996 and 1997 was less than $400 million.) Defense State Secretary Istvan Fodor added that the U.S. presence had important economic benefits for Hungary and facilitated the modernization of the Hungarian army.[114]

During its opening "enhanced dialogue" with NATO in April 1996, Hungary presented its document outlining its position on enlargement issues, steps taken in the field of military reform, and formulated questions regarding accession. When NATO Secretary General Javier Solana visited Budapest on 18–19 April he described the document as "very positive and constructive."[115] In order to neutralize Russian opposition, Defense Minister Keleti proposed that NATO should declare in advance that it would not deploy nuclear weapons on former non-Soviet Warsaw Pact members' territory. Jeno Poda of the parliament's Defense Committee criticized Keleti's proposal as being damaging and one that could lead to mistaken interpretations.[116]

To facilitate the integration process, a NATO integration department was established in the Hungarian defense ministry on 1 June 1996 headed by Ambassador Istvan Gyarmati.[117] Although Hungarian politicians pursued NATO integration, public support was less enthusiastic. In a public opinion poll conducted in April 1996, 48 percent supported joining the European Union, while 18 percent did not feel that membership was so important. In contrast, when asked about NATO only 38 percent supported membership, while 27 percent had a negative opinion.[118] Internal surveys of the Hungarian army also indicated that opinion was divided. Whereas a 1990 survey of regular soldiers indicated that 63.5 percent did not see NATO membership as a good security solution for Hungary, 73 percent in 1993 believed that NATO was a good solution. This declined to 66.5 percent in 1994.[119]

DEFENSE BUDGET AND FORCE MODERNIZATION

Parliamentary Defense Committee MP retired General Bela Gyuricza (Fidesz) noted that the army needed to increase its budget to Ft160 billion in 1996, from 1.4 percent to 2.0 or 2.5 percent of GDP. He added that because of financial shortages the "Hungarian Army is the world's most expensive, because the soldiers are conscripted for 12 months, but only participate in training for four to six days a month . . . and the rest of the time they stay in the barracks."[120] When Col. Istvan Demeter, chief of the HDF's training department, noted that conscripts were being discharged untrained because the HDF did not have enough money for exercises and the physical state of the soldiers was rather poor,[121] Defense Minister Keleti agreed adding that "we need a serious budget to finance the training of our conscripts and these funds are not available."[122]

At roughly the same time, BG Nandor Gruber of the defense ministry Defense Management Department noted that the 1998 army reform goal of a 60,000-strong army could not be maintained because the 1995 parliamentary decision to allocate 2.5 percent of GDP to the armed forces had never been carried out.[123] In fact, the government Cabinet decided to accelerate the cuts because of continued financial constraints. Hence, by January 1997 the HDF had shrunk to 55,730, from 73,660 when Keleti took over, and with the February 1997 intake of conscripts would reduce the conscription time to nine months from twelve months, requiring more resources. The new training system will cost Ft1.2 billion, but only Ft700 million was made available and would only pay for the accommodation of soldiers and the most essential requirements for training.[124]

The 1995 and 1996 defense budget of Ft77 and Ft87 billion, respectively, was increased to Ft96.8 billion in 1997, which in real terms represented no change due to inflation (see table 4.3). As a result, Army Commander LTG Ferenc Vegh noted that the army could *not* comply with the parliamentary decision on restructuring, which requires 15–20 percent of the budget to be spent on development. Restructuring would again have to be put on hold.[125]

The 1998 defense budget requirement requested by the defense ministry was Ft160 billion. While most government and opposition deputies on the parliament's Defense Committee found the figure "reasonable," they argued that it would be "difficult to meet."[126] After the announcement in June 1997 that Hungary would get an invitation to join NATO at Madrid, budget discussions revolved about maintaining the real value of the Ft100 billion 1997 budget, so it would be Ft118 billion plus the Ft8 to 10 billion for NATO admission, totaling Ft130–135 billion.[127]

Although the Conventional Forces in Europe (CFE) allowed the Hungarian Defense Forces to maintain 100,000 troops, by the end of 1997, they comprised only 53,700 troops. Between 1992 and 1997, the number of main battle tanks (MBT) declined from 1,345 to 797, which was below the CFE limit of 835. Armored military vehicles also remained comfortably below the 1,700 CFE limit, having dropped from 1,731 to 1,300 between 1992 and 1997. Artillery pieces had been reduced from 1,047 to the 840 allowed by CFE. CFE limits of 108 for helicopters stood at 59.[128]

Because the army reorganization plan called for troops and equipment to be more concentrated, this required the redeployment of personnel. Storing the HDF's surplus military equipment required over four hundred unused facilities that needed to be guarded by conscripts.[129] The army also would need to reduce the number of military units from 110 locations in 1996 to 81 in 1997. This task required expenditure of Ft150 million, but did reduce the manpower necessary for guard duty.[130] Hence, of the Ft4 billion available for development in the 1997 draft defense budget, most went toward housing; Ft1.69 billion for construction and Ft600 million for housing support for army employees.[131] Two further challenges resulting from the reorganization involved unsolved housing problems and disruption of families

Table 4.3.   Hungarian Defense Budgets

Year	Forints (Ft) billion	US ($) million	% of GDP	GDP (Ft) b	GDP US ($) b
1988	—	—	3.5	—	—
1989	45.5	581.28	2.8	1625	20.76
1990	52.9	613.46	2.5	2080	24.54
1991	55.4	539.83	2.4	2308	22.49
1992	61.7	562.23	2.2	2805	25.56
1993	64.0	516.64	1.9	3368	27.19
1994	66.5	457.65	1.7	3912	26.92
1995	77.0	404.21	1.48	5203	27.31
1996	87.0	397.63	1.32	6604	30.18
1997	96.8	389.13	1.4	6914	27.80
1998	122.0	429.62	1.5	8133	28.64
1999	164.5	509.69	1.6	10281	31.86
2000	192.3	510.98	1.51	12735	33.84
2001	223.6	623.11	1.6	13975	38.94
2002	261.4	731.92	1.6	—	—

Sources: Data compiled from "Hungarian Economy," *Encyclopedia of Hungarian Economy, 1999* (5th edition); www.hungeco.com/economy.html (2 July 2002); "Hungary: Exchange Rate as of May 31, 2002," International Monetary Fund, 2002; and www.imf.org/external/np/tre/tad/exportal.cfm?memberKey1= 415@date1key=2002 (1 July 2002).

because the spouses of relocated officers and warrant officers could not find new jobs.[132]

General Jozsef Hollo of the HDF personnel department noted the effect of reductions on the officer corps. In early 1997 the HDF comprised 8,600 officers and 8,100 junior officers, representing an overall reduction of 1,400 and 1,000, respectively, during 1996.[133] Since nine-month conscripts could not be expected to become tank drivers and artillerymen, the HDF began initial recruitment of 2,000 contract soldiers for posts in 4 towns in 1996 and another 2,000 in 1997 with projected increases of 500 per year thereafter. In early 1997, 4 "professional battalions" consisting of 500 troops each were operating in Tata, Hodmezovasarhely, Debrecen, and Szolnok in eastern Hungary.[134]

During 1996 and 1997 modernization of the ground forces proceeded. By early 1997 all of the one hundred T-72 tanks with their repair kits from Belarus had arrived in Hungary, and LTG Ferenc Vegh noted publicly "it does rub me the wrong way to the extent that purchasing tanks is not the highest priority."[135] He added that when the decision was made that he had not been in his present position, so his opinion was not sought. Also in partial repayment of Russia's $1.8 billion debt, Hungary received $320 million worth of BTR-80 armored transports ($240 million for the HDF and $80 million for the Border Guard).[136] In November 1997, Russia agreed to repay

$250 million of its outstanding debt of roughly $600 million to Hungary, one-half of which was to be in the form of military goods, mainly armored vehicles and parts for the army and Niva jeeps for the Border Guard.[137]

Though Hungary's air force had been modernized with twenty-eight MiG-29s (including six trainers) in return for Russian debt in 1993, Defense Minister Keleti noted that air force modernization lagged behind the land forces.[138] LTG Nandor Hollosi noted that the air force inventory comprised 150 aircraft in 1986. The first reorganization took place in 1989 when the air force budget was reduced 30 percent, forty MiG-21s were withdrawn, and fifty pilots were dismissed. The second reorganization in early 1997 resulted in the withdrawal of another sixty-four aircraft—twelve Su-22s (purchased in 1983), eleven MiG-23s (purchased in 1979), and forty-one MiG-21s—from operation, without reducing the number of pilots.[139]

During 1997 the Hungarian Air Force maintained roughly fifty combat aircraft (twenty-eight MiG-29s, twenty-two MiG-21s; plus ten in "repair reserve"), though Gyorgy Keleti argued the withdrawn MiG-21s "can be put back into service . . . in a military threat."[140] All the MiG-21s, though, would require large-scale overhaul in 1999 and 2000 at a cost of Ft1.5 billion but would add 850 flight hours of service. After the turn of the century Hungary planned to replace its MiG-21s with a new air wing of thirty aircraft.[141] Despite the decline in number of aircraft from 150 in 1986 to 50 in 1997, air force Chief of Staff Attila Kositzky intended to maintain 600 pilots and copilots.[142]

The cost of maintaining and operating the twenty-eight MiG-29 aircraft is substantial. For example, to operate a MiG-29 costs Ft1.391 million plus Ft142 thousand for fuel (e.g., roughly $6,000) per hour.[143] As a result, Colonel Imre Balogh noted that between 1989 and 1996, fighter pilot flying time decreased from 95 to 53 hours per year and helicopter pilots, from 120 to 43 hours. The number of fighter aircraft pilots declined to 79. In 1997 flight time was reduced further to 35–55 hours per year.[144] In addition, to maintain aircraft operations a continued supply of spare parts is essential, but some of the MiG-29s stationed at Kecskemet were grounded for lack of spare parts and remained parked in the open on the tarmac because of lack of hangars.[145]

Defense State Secretary Istvan Fodor noted that Germany's earlier supply of twenty Mi-24 HIND helicopters remained unused for lack of money to refurbish them, adding that even though Germany was willing to supply new equipment free of charge, "it would be too costly to install and put them into operation."[146] Of the roughly sixty helicopters in Hungary's inventory, seventeen Mi-8s and Mi-17 troop carriers were equipped with IFF systems removed from decommissioned MiG-21s during 1997.[147]

Two important additions involved the purchase of low-altitude air defense equipment to be used to defend against helicopters and aircraft attacking army troops in combat. On 25 September 1995 Hungary decided to purchase new radars to replace the existing radars for monitoring civilian and military airspace. Though the radar purchase, which would cost Ft12–13 billion over a ten-year period, was to come from a special budget source, the finance ministry decided that it would *not* assume the expense.[148] Also in late February 1997 a tender worth about Ft18 billion was awarded to the French firm Matra Defense S.A. with the first Mistral-2 air defense missiles delivered in May 1998.[149] In sum, unfunded acquisitions would provide a slow-ticking budgetary time bomb.

The continued build-down of the armed forces and defense budget had a catastrophic impact on Hungary's defense industry. Since most Hungarian acquisitions had been Russian debt offsets, Belorussian/French acquisitions, or German contributions, defense budget allocations for military equipment from Hungarian industry declined from 25 percent in 1988 to 4.5 percent in 1997. Of the Ft77 billion defense budget in 1995, only Ft1.3 billion was spent on military technology modernization. In 1996, it comprised only Ft2.3 billion of the Ft87 billion budget; and declined to Ft2 billion of the 1997 Ft96.8 billion defense budget.[150]

NATO: WEAK SUPPORT AND REFERENDUM CHALLENGE

Hungarian defense expenditures of roughly 1.4 percent of GDP were the lowest of all its central European neighbors (to include Poland, the Czech Republic, and Slovakia) and suggested popular and governmental unwillingness to sustain defense burdens. Even though Bela Gyuricza and Ferenc Vegh continued to stress that Hungary's Ft96.8 billion 1997 defense budget was simply inadequate, the government and defense ministry were either unwilling and/or unable to develop sufficient public understanding of the costs of NATO integration and failed (some might argue did not even attempt) to develop parliamentary support for adequate defense budget levels.

Even the Hungarian Atlantic Council made no effort to prepare the public for the costs of NATO membership when it portrayed an unrealistic underestimation that assumed *no* increase in defense expenditures! Its study estimated that Hungary would incur a direct NATO-membership-related expenditure of Ft4.4 to Ft5.5 billion ($24.5 to $31.0 million) over thirteen years. To this it added the effective contribution of NATO member states of similar size at Ft3.5 billion a year ($20 million). Amazingly, it noted that "taking into account the size of Hungary's annual defense budget—Ft90 to Ft100 billion—the 'direct cost' of joining NATO—4.3 to 5.5 billion—appears to be bearable . . . [suggesting] this amount could even be 'managed' out of a Ft90 to 100 billion annual defense budget, *without* [emphasis added] any additional expenditures and without burdening the population."[151] In other words, NATO is a free ride.

Parliamentary Defense Committee Chairman Imre Mecs (AFD) also noted that joining NATO would *not* be expensive; that defense expenditures would increase only 15–20 percent. Mecs added that if Hungary did not join NATO a similar proportion would be needed anyway as the Hungarian forces had not been adequately upgraded over the past fifteen years. He noted that "NATO compatibility will *not* require huge sacrifices, due to the distribution of tasks among the member states which may even lead to lower costs in the long run."[152]

In marked contrast to the assessments of the Hungarian Atlantic Council and Imre Mecs, U.S. Defense Secretary William Cohen emphasized in Budapest in July 1997 that improving military training and devising defense programs would be expensive, though adding that the expenses were affordable. The U.S. defense secretary stressed that NATO expected Hungary to increase its military budget to 2.1–2.2 percent of GDP.[153] Hungarian Colonel Laszlo Nagy also provided an independent assessment that was more consistent with Defense Secretary Cohen's estimation that NATO integration would be costly. Drawing on work prepared by the Defense Ministry Integration Secretariat, Nagy estimated that over the next thirteen years the Hungarian contribution to the direct costs of NATO accession would be Ft108–144 billion, and the cost of the transformation of the army would be Ft360–484 billion. Nagy also added the significant caveat that if Hungary did not join NATO that the Hungarian army would have to be transformed at an even more urgent pace. Nagy estimated Hungary's yearly cost of accession to be Ft36–48 billion, requiring an increase of approximately 37–50 percent in the 1997 defense budget.[154]

Public opinion polls of Hungarian society consistently indicated a lack of support for defense, and the Hungarian government failed to prepare the public to shoulder this burden. While most Hungarians (85 percent) expected that joining NATO would increase the amount needed for defense, a majority (58 percent) *opposed* such an increase. In marked contrast, while 88 percent of Polish society believed that NATO membership would require increases in defense expenditures, 55 percent would *support* them.[155] If an increase in Hungarian military spending came at the expense of social needs, opposition was overwhelming: 86 percent in 1995; 87 percent in 1996; and 88 percent in 1997![156]

Of the three Partnership for Peace partners invited to join NATO in 1997, only Hungary required a referendum. During the MDF government of 1990–1994, when the parliament discussed the basic principles of Hungarian security policy and NATO and European Union membership, Gyula Horn of the opposition HSP suggested that "following the debate and decision of the Parliament [on] . . . a matter of such importance . . . The society should decide in the framework of a referendum whether it wishes the accession to a new alliance system NATO or not."[157] Hence, when the HSP formed a government with the AFD after the 1994 elections, it promised that after the conclusion of talks about NATO membership it would ask the opinion of the population. Initially Prime Minister Horn planned a consultative—nonbonding—referendum and criticized the Fidesz leaders who wanted the referendum to be legally binding.[158] In the end, though, when the decision was announced on 28 August 1997 to hold the referendum on 16 November it would be legally binding because AFD coalition partner Deputy Prime Minister and Interior Minister Gabor Kuncze "regard[ed] NATO membership as a matter of extraordinary importance [and] for this reason they accept the referendum as binding."[159]

Indeed, for all the above noted reasons, such a referendum was probably necessary to validate the Hungarian government's bid, but it did create some cause for concern in Brussels and resonated on the broader NATO ratification process. The Hungarian referendum question to be answered with a simple "yes" or "no" simply stated: "Do you agree that the Republic of Hungary should guarantee its security by joining NATO?" The question masked Hungary's membership obligations and burdens with no hint of what Hungary needed to do for NATO and intended to elicit a positive response: that Hungary would get its security guaranteed by NATO.[160]

Although Prime Minister Horn underscored the Hungarian government had no reason to fear binding referendums and emphasized his absolute confidence that a large proportion of voters would favor accession,[161] a number of potholes lined the road up to 16 November 1997, as well as after. One was the U.S. Senate's plan to ratify the NATO enlargement decision sometime during the spring of 1998. The Hungarian referendum would be an important consideration in the U.S. Senate ratification, as well as with many other NATO members.[162] Another potential pitfall related to popular turnout. Even if high support was registered for NATO, low turnout coupled with Hungary's low defense expenditures could certainly raise serious questions in the U.S. Senate about the depth of Hungary's commitment. According to Hungarian public opinion data conducted in early August 1997 at the direction of the Ministry of Foreign Affairs, 43 percent said they would "surely" (and 28 percent said "probably") participate in the referendum; 15

percent said they would "surely" (and 8 percent "probably") *not* participate. Of the 43 percent who would surely participate, 76 percent supported NATO accession.[163]

The Hungarian government needed to generate a high popular turnout and to demonstrate broad Hungarian support for NATO. If Hungarian support proved to be weak, it would likely have repercussions on U.S. Senate ratification. Hungary was the most vulnerable of the three candidates for NATO membership because of its low social support for NATO, lack of understanding of NATO obligations, low defense budgets, and unknown implications of its referendum. The U.S. Senate needed to be convinced that new members will be "producers" of security, and Hungary needed to overcome the image that it would only be a "consumer" of security. In the end, the government campaign was "successful." Though less than half (49.6 percent) of the eligible voters went to the polls (far less than the 65 percent and 69 percent first-round turnouts in the general elections in 1990 and 1994), the 16 November 1997 referendum was considered binding because the recently changed Hungarian electoral law required that only 2,000,001 people must vote yes. At the referendum 85 percent (3.3 million people) supported NATO membership.[164]

HUNGARIAN DEFENSE FORCES AFTER MADRID

Hungary's senior military leadership understood that military training and force modernization required significant attention and development to meet NATO standards. It remained unclear whether Hungarian politicians really understood or accepted this. Shortly after the July 1997 Madrid summit Col. Gen. Ferenc Vegh expressed frustration with civilian politicians. General Vegh acknowledged Hungary's military deficiencies, noted that NATO has a five-year planning cycle, and in apparent frustration added that Hungary would:

> probably be unable to answer [NATO's] question on the army budget and army equipment in the year 2001 because . . . a transparent planning system has not yet emerged.
>
> It is the task of politicians and financial experts to plan and distribute the budget. *This means both civilian control and civilian responsibility* [emphasis added] . . .
>
> Budget planning is necessary not only annually, but also for the medium- and long-term. This is how it is done in every NATO-member country, and also the other two new candidates. The Czech Republic and

Poland have already stipulated the percentage of their GDP for defense expenditures.

> We need a bigger budget not because of NATO membership. We have been spending on material expenses and technological development below the norms for many years. . . . If we cannot increase the material expenses, we will not be able to fulfill further NATO requirements! NATO does not need untrained soldiers![165]

Ferenc Vegh also specifically outlined the HDF's three defense tasks. First, NATO's common defense, as stipulated by Article 5, required Hungary to develop a national support capability and to establish a well-organized professional army because its (now nine-month) conscripts could not be sent abroad. Second, the HDF needed a peace support and crisis-management capability. Third, the HDF had to guarantee Hungary's defense and territorial integrity[166] A few months later Vegh reiterated that conscripts would never be sent abroad, and that Hungary would offer its technical battalion currently serving in Bosnia, and a brigade, composed of members of the present Tata, Hodmezovasarhely, and Debrecen brigades, to NATO's reaction forces.[167]

In order to deal with NATO membership, the HDF established a new command structure on 1 September 1997 (completed on 31 December), which included an armed forces staff headed by Col. Gen. Ferenc Vegh and separate staffs for the land forces in Szekesfehervar under LTG Ambrus Preininger, and air force in Veszprem under MG Attila Kositzky.[168] NATO-compatible units composed of professional and limited-contract soldiers were set up in four Hungarian cities. Since HDF conscript service time had been reduced to nine months, conscripts could not be used for international tasks;[169] hence, by 2010 the HDF was to include only professional and limited-contract soldiers.

The first call-up of 7,100 nine-month conscripts took place on 17–18 November 1997. Though these conscripts were subject to more "intensive" training—constituting six weeks of basic and six weeks of technical training at the four training centers (Tapolca, Szombathely, Kalocsa, and Szabadszallas) before being sent to their units[170]—significant conscript problems remained evident. According to Colonel Lajos Erdelyi, 35 to 40 percent of the 7,500 conscripts in the February 1998 call-up were unsuitable for military service because of health.[171] LTG Ambrus Preininger added that drugs were a "time-bomb"—32.6 percent of Hungarian conscripts used drugs while on leave on weekends.[172] In addition, Hungary's demographics suggest that by the year 2003–2005 the available intake will decline to roughly 50,000 from the 90,000 available in 1998. This factor alone will undoubtedly have a significant impact upon the future HDF and its professional composition.[173]

MDF Parliamentary Defense Committee member Jeno Poda expressed concern about the HDF's ability to implement its tasks, noting that "with the exception of a few major Army exercises, there is no training whatsoever in most garrisons."[174] LTG Preininger noted that despite the lack of adequate conditions, the land forces fulfilled their tasks during 1997, though they failed to improve the living conditions of their personnel.[175]

But without social support and adequate resources, Hungary's ability to build a more professional force would prove difficult. Of the desired initial 4,500 contract soldiers stipulated by the defense ministry for 1997, BG Istvan Szekeres noted that by August 1997 only 2,200 positions had been filled. Most were unemployed youth with only an eighth-grade education; very few had secondary school certificates.[176] Many youth from western Hungary preferred to seek employment in Austria; while many of those who did join from the east, quit after three months in order to qualify for unemployment benefits.[177] In sum, the program for building contract soldiers was faltering.

The professional soldiers were also disgruntled. Though Defense Minister Keleti had promised them a wage increase of 23.5 percent against a planned inflation rate of 13.5 percent, the Soldiers' Trade Union organized a series of demonstrations demanding a 41 percent increase.[178] After eighty officers marched to the finance ministry in protest on 9 October 1997, Prime Minister Gyula Horn, called the demonstration irresponsible and criticized their wage demands.[179]

Another change in the composition of the professional military has been the buildup of women; numbering 2,416, or slightly more than 10 percent in mid-1998. Most women were noncommissioned officers, while 461 were officers (mostly captains and majors). Roughly 25 percent were employed as doctors, nurses, and other medical staff. The first women's unit of twenty-seven soldiers completed financial, telecommunications, and technical studies at Bolyai Janos Military Technical College in August 1998.[180]

Hungary's forces already felt their limitations when they deployed their noncombat Engineering Battalion of 416 troops to IFOR in Croatia in late January 1996. Even though the Engineering Battalion had been reduced to 310 troops during SFOR in mid-1997, when fresh troops were needed, Hungary found it difficult to rotate them and maintain the reduced presence.[181] In addition to their 310 troops in SFOR, another 105 Hungarian solders were serving in Cyprus in an Austrian-Hungarian-Slovenian battalion by April 1998.[182] The Hungarian armed forces faced another challenge with the new requirement to establish a Hungarian-Romanian battalion of five hundred troops each as a peacekeeping unit by autumn 1998.[183] In addition, Hungary was making efforts to develop a combined brigade with Italy and Slovenia, each contributing one battalion.[184] This effort marked the first time an integrated military structure would bring together a NATO member with non-NATO members. In order to facilitate this trilateral cooperation, the three defense ministers had been meeting twice per year since 1996.

After the NATO invitation had been issued in Madrid in July 1997, Hungary began to reassess its needed defense acquisitions. Col. Gen. Ferenc Vegh noted that with NATO membership Hungary would need to establish its Airspace Sovereignty Operations Center (ASOC) in Veszprem by mid-1998. Zoltan Martinusz added that the radars were needed not only because of NATO membership, but because existing radars were "obsolete to the point of being useless."[185] The center was to be equipped with two radars to be operating by the date of NATO accession.[186] The first phase was to begin in February 1998; the United States contributed the majority of the cost ($6.5 million) and the Hungarian defense ministry spent Ft130 million ($513,000).[187]

One controversy that arose over the radars was the manner in which the defense ministry's Procurement Office had handled the tender. Imre Mecs, chairman of the Parliamentary Defense Committee and an electronics engineer, criticized the decision to issue a Ft500 million tender to Siemens over four other companies. Mecs asked Defense Minister Keleti to have an independent committee reassess the bid and request Siemens to provide references from the *Bundeswehr*.[188] In response, Keleti noted on 17 March that no decision had been made yet and that he would inquire about Siemens from the German *Bundeswehr*.[189] On 22 June 1998 the tender was declared null and void.

Istvan Gyarmati, Deputy State Secretary for Defense Policy, outlined three tasks to be resolved in military resource management for 1998. First, officers and warrant officers needed a clear vision of their career and needed to understand criteria for promotion. Second, training needed to be "radically strengthened," including improving everyone's physical condition. Third, Hungary needed to develop a national military strategy, doctrines for the various forces and weapons, and new regulations.[190]

At the same time, Hungary's NATO accession negotiations proceeded in September and October 1997 on political, military and defense, and financial issues. After the financial round in October, the Alliance proposed that Hungary pay 0.65 percent of NATO's budget—Ft2.3 billion or $11.7 million—in annual dues.[191] As a result of the NATO coordination talks, Hungarian leaders began concluding that early acquisition of fighter aircraft might *not* be necessary. Foreign Minister Kovacs noted that it was a higher priority to "buy modern radar equipment than to buy fighter aircraft."[192] In fact, NATO asked Hungary to postpone its tender for purchasing radars to ensure that the three new members would be linked in a coordinated way with the air defense system.[193] In addition, Ferenc Vegh concluded that those pilots who were serving in units that were cooperating with NATO would need to have their flight time increased from the low of thirty hours to eighty hours.[194]

Another requirement for Hungary's NATO integration is the need to establish a crisis-management mechanism to enhance civil-military cooperation and coordination to deal with new security challenges. Vegh noted that "military forces alone cannot handle any kind of crisis in the future. Military forces will be only part of wider crisis management systems. These mechanisms, in which both civilian and military components are able to handle a crisis, should be set up."[195] In February 1998 Hungary participated in a NATO crisis-management exercise CMX-98 in which more than four thousand workers in fifty organizations participated with the military.[196]

Though no decision would be made until Hungary joined NATO, Defense Minister Keleti expressed his preference to have Hungary participate in NATO's southern command at Naples.[197] In addition, Keleti noted in January 1998 that eighty Hungarian officers would be posted to Naples, Mons, and Brussels in April.[198] A few months later Keleti noted that NATO integration would require more English-language speakers trained at NATO's standard STANAG 6001 at higher levels and in combat units.[199] By the spring of 1998, almost 1,500 members of the professional staff of the Hungarian army spoke English and 150 spoke French. Starting in 1998 at least six hundred individuals would study English.[200]

THE CHALLENGE OF IRAQ

On 10 February 1998 when U.S. Secretary of State Madeleine Albright asked Foreign Minister Kovacs for support against Iraq, Kovacs noted that he could not give a final answer without concrete authorization from government or parliament, but he gave his personal opinion that he thought Hungary would act no differently than during the Gulf War in 1991.[201] Two days later (on 12 February) the Hungarian government decided to support a military operation if a political solution became impossible.

The government and parliament then moved quickly. On 17 February the parliament's foreign affairs and defense committees pledged unanimous support for the government proposal to allow countries participating in the anti-Iraq action to use Hungary's airspace and designated airfields, and for Hungary to send fifty military doctors to the region.[202] On 18 February Ferenc Vegh noted that preparations had commenced at three airfields (Papa,

Kecskemet, and Taszar);[203] and on 19 February, the government approved Ft550 million from the budget reserve; of which Ft270 million would be used to purchase chemical neutralizers, fire trucks, satellite telecommunications, and container loaders, and Ft250 million for preparing and sending the medical team.[204] After UN Secretary General Kofi Annan's negotiations averted the crisis, Keleti stopped acquisitions on 24 February 1998 after Ft120 million had been expended.[205]

In summary, the Horn government mandate failed to adopt a new constitution, but did successfully implement a number of amendments to ameliorate earlier deficiencies such as the Constitutional Court's powers. The Parliamentary Defense Committee had continuity and opposition representation, though limitations were evident in its lack of oversight of the T-72 acquisition from Belarus and in deployment of MiG-29s to Poland for a PFP exercise. It did approve Hungary's contribution to SFOR. When retired Colonel Gyorgy Keleti became defense minister he reorganized the General Staff, providing it with more authority, decided not to merge it with the defense ministry, redivided the two top army positions (HDF commander and chief of staff), and downsized the defense ministry, bringing into it many of his former military officers and raising questions about "civilian control" of the military. Ministry shortcomings remained evident in mid- to long-term budget planning and transparency, military and defense planning, acquisitions, and in personnel policy and legal regulations. Although Keleti had a very strong position within the Hungarian Socialist Party, he never fought for more defense resources, and the budget continued its downward spiral from 1.7 percent of GDP to 1.5 percent during his mandate. Procurements were made in low-altitude radars for air defense and Mistral-2 air defense missiles, but military housing remained a serious problem. Reductions in the armed forces were achieved mainly by reducing conscripts by 40 percent, from 51,560 to 33,000, and by reducing their service time to nine months. Even though Hungarian law does not permit the deployment of conscripts abroad, nine months training virtually guaranteed this. Limitations in the Hungarian forces quickly became more evident after January 1996 with the need to maintain a Hungarian IFOR/SFOR deployment by sustaining a smaller 310-troop presence through rotations. Hungarian support for a potential operation against Iraq included the use of airspace and airfields and sending military doctors to the region. Although career soldiers were only reduced by 6.3 percent, from 22,100 to 20,700, they were unhappy over the May 1996 Legal Status Bill, their pay, and professional training. As military stocks and spares had been consumed, the armed forces participated in fewer military exercises, readiness levels declined, and morale dropped.

5

1998 Parliamentary Elections: Center-Right Government Returns

In view of the impending elections on 10 and 24 May the parliament closed the last session of its four-year cycle on 16 March 1998. In the first round (10 May) of the elections, the voter turnout of 56.25 percent was the lowest in the history of post–Cold War Hungary, and six political parties passed the 5 percent threshold. The HSP received 32.25 percent of the vote; Fidesz-MPP 28.19 percent; Jozsef Torgyan's FKGP 13.77 percent; Gabor Kuncze's AFD 7.88 percent; and Istvan Csurka's right-wing Hungarian Life and Justice (MIEP) Party, 5.55 percent. After the first round, the FKGP reached accord with the MDF and then with Fidesz-MPP to withdraw candidates in relevant constituencies; the AFD did the same with the HSP.

In marked contrast to the first round, the 24 May turnout at 57 percent was the highest in second-round parliamentary elections since 1989 (45.5 percent in 1990 and 55 percent in 1994); and six parties were able to pass the 5 percent threshold (see table 5.1). In effect the Hungarian Socialist Party–led government of Prime Minister Gyula Horn fell from power when its 1994 absolute majority of 209 (of 386) seats declined to 134 in Parliament. Fidesz-Civic Party captured 148 seats (from only 20 in 1994), and its leader Viktor Orban became the new Hungarian prime minister (see figure 5.1). The defense ministry went to the FKGP. While the defense minister and political state secretary posts went to Janos Szabo and Janos Homoki of the Smallholders Party, Tamas Wachsler of Fidesz occupied the administrative state secretary position.

The new parliament established twenty-two committees of which three were new (i.e., tourism, youth and sports, and regional development). The constitution-drafting committee was dissolved. Former Defense Minister Gyorgy Keleti (HSP) chaired the fifteen-member National Security Committee; Zsolt Lanyi (FKGP) chaired the twenty-one-member

Table 5.1.   Hungarian Parliamentary Elections: 10 and 24 May 1998

Political Party	Seats
Government (N = 213)	
Federation of Young Democrats-Hungarian Civic Party (Fidesz-MPP)	148
Independent Smallholders and Citizens Party (FKGP)	48
Hungarian Democratic Forum (MDF)	17
Opposition (N = 172)	
Hungarian Socialist Party (HSP)	134
Alliance of Free Democrats (AFD)	24
Hungarian Life and Justice Party (MIEP)	14
Total	386

Defense Committee; and Istvan Szent-Ivanyi (AFD), the twenty-six-member Foreign Affairs Committee.[1]

PRIME MINISTER CHANCELLERY AND DEFENSE REFORM

Viktor Orban decided to elevate security and defense policy to the prime minister's office, where he attempted to create a super-chancellery along the German model. This intention was reflected in the doubling of the chancellery budget for 1999 to Ft2.335 billion and the increase of at least two

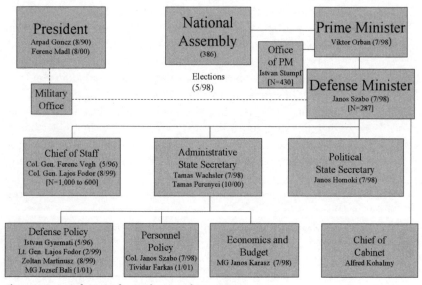

Figure 5.1.   Orban/Szabo Defense Reform, 1998

hundred more staff than were present in Gyula Horn's chancellery.[2] Retired General Bela Gyuricza became the secretary of state for Security and Defense Policy. Perhaps it was for the purpose of gaining personal power that Orban permitted his junior FKGP coalition partner, with little experience or interest in defense, to assume control of the defense ministry while Hungary was preparing for NATO integration.

The new government program contained a chapter called the Security Policy and Defense, which noted as an urgent task the need to integrate the army general staff into the defense ministry and end the parallel structure that had been in existence over the past eight years. Though both Gyorgy Keleti (HSP) and Defense Committee Deputy Chairman Imre Mecs (AFD) agreed with the objective, they believed that "the authority and tasks of the Chief of General Staff needed to be precisely defined."[3]

In one of his first interviews Janos Szabo (FKGP) admitted "surprise" when Jozsef Torgyan summoned him to Parliament to offer the defense portfolio "because I had not much dealt with national defense before."[4] Shortly after his first briefing about the state of the army, Szabo noted that he was "a bit shocked [concluding that] the Army's situation is far from rosy."[5] A few weeks later Defense Minister Szabo noted that over the next four to five years it would be necessary to increase the number of warrant officers and contract soldiers from the four thousand then in service to eighteen thousand; and to retain conscripts, but reduce their numbers. Szabo also noted that the January 1999 wage scale would include a 16 percent nominal increase (really 5 percent because of 11 percent inflation), and that he would introduce a new housing concept to attract soldiers.[6] Defense Political State Secretary Janos Homoki affirmed that though the budget would increase 0.1 percent in real terms, that actually more would be needed because professional soldiers were to receive higher wages starting in 1999.[7]

Problems in the air force were immediately apparent. On 23 July 1998, one of Hungary's twenty-eight MiG-29s crashed in Kecskemet. The pilot was killed and a military committee inquiry established that the crash was not due to a technical defect.[8] Claiming that poor operating conditions could no longer be tolerated, Col. Gen. Ferenc Vegh ordered a full examination of the air force, adding that the budget should be increased by at least Ft50 billion. Vegh admitted that due to spare parts problems and lack of money for repairs only fourteen of the twenty-seven MiG-29s (55 percent) could be deployed, and that the twenty Mi-24 helicopters that Hungary had received from Germany could *not* be put into service because Hungary did not have money for repair costs.[9]

Hungarian fighter pilots noted that fifty to sixty hours of fight training were significantly insufficient for maintaining combat readiness. In

Kecskemet only fourteen of the forty MiG-29 pilots were capable of operating combat tasks day or night and only twelve were capable of operating with NATO.[10] In early November 1998 Hungarian air force chief MG Attila Kositzky told the Parliamentary Defense Committee that Hungary could *not* fulfill its commitment to have its twelve NATO-designated pilots fly at least 100 to 120 hours to maintain combat readiness because the lack of spare parts limited operational time for the combat aircraft.[11]

The air force chief of staff's December 1998 report on the state of the air force outlined three options. The best case assumed Ft30 billion for the air force, which meant pilots would get a salary increase of Ft5 billion and would fly 100–120 hours instead of the current 50–60. The second option, involving modest financial constraints, would *only* guarantee increased flight time to the MiG-29 pilots. The worst case assumed no budget increase and would prepare only the eight MiG-29s and crews pledged to NATO to be operational.[12] Since the air force's share of the 1999 defense budget (Ft16.3 billion from Ft16.0 billion in 1998) meant the worst case would become reality, the air force had to modify its past practice of ensuring at least twenty to thirty hours a year for everyone and pilots would be dismissed. Since only pilots who fly at least one hundred hours should fly for safety reasons; now only one-third of Hungary's 150 combat pilots (including the 12 designated for NATO) would fly, while the others would carry out ground services. In sum, now only seventeen pilots (one-half the required number) pilots would fly the twenty-seven MiG-29s, of which more than half were not operationally safe.[13]

Problems in the land forces also continued to deepen. Most vehicles were clearly outdated. The nineteen thousand wheeled vehicles in operation in the fall of 1998 represented a decline from thirty thousand a decade earlier because many had been cannibalized for spare parts. The modernization program planned for 2005 required Ft50 billion, but available finances would only permit the purchase of one hundred vehicles in 1999, with plans for acquiring only two to three hundred per year thereafter.[14]

By mid-September 1998 apparently Janos Szabo was beginning to have different thoughts about the Hungarian army's future in light of Hungary's impending NATO accession. The new defense minister noted that one of the major modernization tasks was to make troops deployable in twelve to twenty-four hours, rather than the current six days. Szabo referred to a RAND assessment that supported smaller and more mobile forces and recommended cutting the number of Hungary's fighter aircraft and increasing combat helicopters.[15]

NATO INTEGRATION CHALLENGES

Of the forty-eight tasks NATO recommended the Hungarian army complete, Hungary agreed to fulfill forty-four over a five-year period. As the 12 March 1999 date of NATO accession rapidly approached, Hungary could only meet 60 percent of its membership commitments. Hungary made commitments to implement thirteen military-technical projects by accession—to include coordinated air defense and telecommunications systems, installing rapid reaction forces, improving language skills, and harmonizing laws to enable NATO to send troops in a crisis—and sixteen by the end of 1999.[16] Col. Gen. Ferenc Vegh warned that the new Orban government program to reduce conscription to six months would mean that conscripts could *not* perform certain necessary tasks. As a result, he would need to expand the number of contracted soldiers, but available resources prevented this.[17]

NATO membership also meant the defense ministry would need to accomplish additional tasks. First, it had to rewrite the National Security Strategy and Military Strategy documents *before* Hungary's accession to NATO. Second, it would need to amend the Constitution and Defense Law because of frequent NATO military exercises. During peacetime parliament did not need to authorize troop movements, but during wartime, it would retain its jurisdiction. Third, the General Staff needed to be fused with the defense ministry for compatibility.[18] Fourth, it needed to meet certain military requirements by the time of Hungary's accession to NATO in spring 1999. Ferenc Vegh noted that Hungary had to fulfill eleven requirements, which included joining the integrated air defense system, securing classified materials and documents, preparing Hungary to accommodate NATO (such as upgrading domestic airports),[19] and augmenting staff at NATO commands. Initially the Hungarians assigned a liaison fourteen-man contact team to SHAPE, then to NATO's AFSOUTH Command at Naples. Overall seventy-two officers were to be sent to various European command centers and six to Norfolk, Virginia.[20]

But faltering support for one political commitment raised serious concerns! The leaders of five parliamentary groups (MIEP excepted) initially agreed to amend the Hungarian Constitution to allow the government to send troops abroad and receive foreign troops on Hungarian soil. But the HSP and AFD withdrew their support of the amendment on 2 December 1998 because of shaken confidence in the government, and they now wanted to retain the necessary two-thirds rather than simple majority parliamentary endorsement. On 15 December, parliament defeated the constitutional amendment by not meeting the necessary two-thirds (258 votes); only 204 deputies supported the amendment, with 108 abstentions, and 14 opposed.

Prime Minister Orban assured NATO that Hungary's membership intentions remained unchanged, and the parliamentary opposition position was merely a "temporary blip."[21] Foreign State Secretary Zsolt Nemeth noted that the "diplomatic assessment is that this is an awkward situation. The less diplomatic assessment is that the opposition's conduct over this can be viewed as disregard for national interests."[22] U.S. Secretary of State Madeleine Albright referred to the situation as a "bad sign."

In March 1999 the government was still making efforts to find a compromise solution and eradicate this embarrassing situation. A new constitutional amendment still required advance parliamentary approval for Hungarian armed forces participation in peacekeeping missions; however, now supported by only two-thirds of the parliamentary deputies *present* (and not of the 386 total). The constitutional amendment, though, allowed the government to take decisions and then inform parliament of troop movements related to exercises.[23]

1999 CHALLENGES

The new government pegged the 1999 defense budget at Ft164 billion. The defense ministry would have to sell real estate in order to finance some of its development projects. The ministry hoped to raise Ft13 billion by selling three to four hundred properties worth Ft18 billion.[24] In December 1998 the six-party negotiations in parliament accepted the new National Security Strategy draft. Peter Siklosi, chief advisor in the PM's office, noted that though a detailed military strategy would not be formulated until June 1999, it was already clear that the sixty thousand-troop Hungarian armed forces could *not* continue. The government wanted to fuse the army headquarters in the defense ministry by the end of 1999 and to start to reorganize the mixed army comprised of more professional soldiers and six-month conscripts in the year 2000.[25] Defense Minister Szabo noted that the current plan was to have one-half conscripts and professionals by 2005, which meant increasing the number of professionals by one thousand each year.[26]

But this was easier said than done. The professional corps was out of balance. In June 1998, the Hungarian armed forces had fifty-eight generals (twice the necessary number) serving in the forces.[27] The 8,400 officers were top heavy with too many colonels and lieutenant colonels and were slated to decline to 7,000 by 2003.[28] Though 28,000 nine-month conscripts would be called up annually (7,000 every three months), the May 1999 call-up was larger than normal because the army wanted to ensure full staffing and expected that some would not show up (4,500 chose conscientious objection in 1998).[29]

Immediately before accession to NATO, the defense establishment appeared in disarray. On 14 January 1999 Hungary first admitted publicly that it would not comply with NATO force goal targets. The defense ministry announced it would launch an investigation to determine if a professional or administrative mistake had been made when filling out Hungary's NATO Defense Planning Questionnaire (DPQ). On 27 January Defense Minister Janos Szabo announced the dismissal of Land Forces and Air Force Commanders LTG Ambrus Preininger and MG Attila Kozitsky, replacing them with MG Andras Havril and BG Istvan Talla, respectively. Then upon completion of the DPQ investigation (9 February), Defense Minister Szabo relieved Deputy State Secretary Istvan Gyarmati for allegedly submitting a DPQ that contradicted the intentions of the ministry and government.[30] State Secretary of Defense Janos Homoki refused to confirm or deny if Hungary would resubmit the DPQ.

At the same time, NATO preparations were in full gear. In January Hungary established the National Security Supervision for handling and treating NATO classified information. On 8 February, the Airspace Sovereignty Operation Center (ASOC)—which linked Hungary to NATO—opened in Veszprem. On the next day, the Hungarian parliament voted (330 for, 13 opposed, with 1 abstention) to join NATO.

Chief of Staff Ferenc Vegh, though, made it clear that after formal accession on 12 March 1999, Hungary still had an enormous amount of work to accomplish in the fields of air defense and air space control modernization, foreign language instruction, improvements in reconnaissance, and increasing the combat capacity of rapid reaction troops. Vegh specifically noted that of the Ft164 billion 1999 defense budget, only Ft7 billion (less than 5 percent) would be spent on development. Vegh predicted the Hungarian armed force would reach "intellectual capability by 2003 [and] technological . . . by 2013."[31] According to BG Zoltan Szenes, AFSOUTH deputy chief of staff for logistics, Hungary's Rapid Reaction Force's mechanized battalion (Tata) would be ready by the fall of 1999, but that the incorporation of Hungary's main defense corps would take about five years.[32] LTG Lajos Urban, first deputy chief of staff, noted that "the real military work will only start now, and it will be no simpler or easier than the work that we have done."[33]

Defense Minister Szabo noted that the Hungarian army would spend Ft4.5 billion in 2000 to buy new vehicles, that the defense ministry wanted to buy a MiG-29 simulator in 1999, and that he hoped a new fighter aircraft tender would be announced by "2003 at the latest."[34] Hungary's MiG-29s posed two problems. First, its navigation and communication systems were not suitable for carrying out flights or exercises in the original NATO members' airspace under routine circumstances for another two to three years. The aircraft could only provide information for flight control, but were unable to receive NATO's identification code.[35] Second, according to Janos Szabo the Russians were not reliable partners because promised deliveries of MiG-29 spares had not been received.[36]

KOSOVO CHALLENGES

The emerging situation in Kosovo meant that Hungary would need to provide airspace to NATO. When NATO's formal request arrived on 11 October 1998, Viktor Orban convened a special cabinet session to deal with the issue. The government approved the proposal on 13 October and the parliament on 14 October. The Hungarian SFOR unit in Bosnia was also authorized to carry machine guns for protection.

As Kosovo continued to deteriorate, on 18 February 1999 Defense and Foreign Administrative State Secretaries Tamas Wachsler and Janos Herman informed parliamentary groups about the possible need to participate in Kosovo peacekeeping (KFOR), which would require two-thirds vote. As military operations became more likely, Foreign Minister Janos Martonyi noted that Hungary's Taszar base providing logistical support for SFOR in Bosnia would not be expanded, and that the border guard had been augmented by one thousand troops (to now total three thousand). Martonyi noted that in the event of air strikes, NATO aircraft would use Hungary's airspace and would need landing rights, which required parliament to make a decision on 24 March with a 93 percent majority (252 for, 12 against, and 12 abstentions).[37]

When military operations commenced during the evening of 24 March, Viktor Orban publicly noted that Hungary would

> definitely not send soldiers to Yugoslavia, however, we shall ensure technical preparation. NATO has so far not requested, nor will it in the future, armed participation from Hungary and, for that matter, our country could not fulfill such an expectation in view of its special situation.[38]

Hungary's "special situation" is that it was the only NATO country bordering Yugoslavia, and that 300,000 Hungarians live in the Vojvodina.

On 13 April Parliament again approved (263 for, 17 against, and 14 abstentions) to send a group of forty health-epidemiological specialists to Albania.[39] On 18 April a Hungarian medical team of six epidemiologists and four assistants left for Albania to provide medical treatment for Kosovar refugees and to prevent further spread of an epidemic. In total the team comprised thirty-five Hungarian army public health officials who remained in Albania until mid-August 1999.

In early April one tense moment occurred when a Russian-Belorussian convoy of eighty trucks appeared at the Hungarian border in transit to Yugoslavia. After days of negotiation Hungary agreed to permit the convoy to pass through Hungary only under customs and police supervision. Hungary refused to allow five armored vehicles and four of the eight fuel trucks to enter into Yugoslavia because they contained excess fuel, which was embargoed. During the same period, after two Yugoslav MiG-29s flew into Hungarian airspace, the Parliament's Defense Committee authorized the defense ministry to purchase night-vision equipment and bullet-proof vests for Ft2 billion.[40] Tamas Wachsler noted that one of the consequences of these unplanned military activities—and the flood costs—was that the defense budget was further stretched by billions of forints.[41]

Concerned about the more than 300,000 Magyars living in Vojvodina, public opinion polls immediately before the NATO Washington Summit indicated that 54 percent supported (42 opposed) the air strikes; two-thirds felt that "under no conditions should NATO resort to a ground war."[42] For these reasons, the Hungarian Parliament Foreign Affairs Committee reached an agreement on 21 April among the five parties that "Hungary would *not* directly participate militarily in ground operations . . . [and] would *not* make [its] territory available for the action" (emphasis added).[43]

On 27 April Janos Szabo noted that an agreement had been reached with NATO so that twenty aircraft (including eight KC-135 tankers) could take off from Budapest's Ferhegy 1 airport to refuel NATO fighter aircraft engaged in bombing missions in Yugoslavia.[44] Upon return from the Washington Summit, Defense Minister Janos Szabo reported to the Parliament's Defense Committee that "[I]t now seems the Hungarian Government would not, *under any circumstances*, contribute to a ground action."[45]

Prime Minister Viktor Orban noted that Hungary would support NATO's actions against Yugoslavia by supplying military bases and an air corridor to NATO (24 March Parliament decision), but would *not* under any condition provide Hungarian ground troops. Apparently unconvinced and skeptical, the opposition HSP—with 134 of 386 parliamentary seats—wanted to reopen the five-party agreement, providing NATO with airports for supporting military operations against Yugoslavia.[46] Foreign Minister Martonyi responded that the HSP proposal to limit NATO's use of airports was "incomprehensible," while some in Brussels noted "surprise [and] incomprehension."[47]

Viktor Orban, in a novel interpretation of NATO's Article 5, added that he had emphasized to the Washington Summit that "the issue of the Hungarians in Vojvodina is not a Hungarian issue, but an issue concerning NATO, and Milosevic must view it accordingly . . . if the Hungarians are harmed to the slightest extent, there must be an appropriate response."[48]

When the number of NATO air strikes against Yugoslavia increased at the end of April, Prime Minister Orban admitted that Hungary's armed forces had been upgraded and strengthened along the southern border.[49] On 5 May when Defense Minister Szabo announced that twenty-four U.S. F/A-18 Hornet fighters and three A-10s would soon arrive at Taszar, 72 percent of the Hungarian public was opposed to NATO launching air strikes from its soil, but 54 percent felt Hungary would lose prestige if Parliament decided to adopt the HSP proposal to modify the earlier decision to permit NATO use of its airspace.[50] Foreign Minister Martonyi admitted that the HSP was split on the issue of the Yugoslav conflict, and that Hungary's opposition to a ground campaign at the NATO Washington Summit was the reason that the option was *not* included on the agenda.[51]

When the final F-18s had arrived at Taszar on 26 May, NATO announced that the fighters could take part in air strikes against Yugoslavia. Defense Minister Szabo voiced the possibility that the F-18s would not be sent into combat, but added NATO's supreme command would make the decision if they were needed.[52] After the three A-10s arrived, NATO on 28 May launched its first air strikes from Hungary against Yugoslavia.

Laszlo Kovacs, HSP parliamentary floor leader and former foreign minister, proposed that Hungary should ban NATO aircraft from taking off from Hungarian airports to attack Serbian targets.[53] Parliament rejected the proposal on 1 June 1999. Despite the apparent fraying of Hungarian support, on 2 June Parliament approved participation of five Hungarians in NATO's rapid deployment force in Kosovo and *all* six parliamentary groups, including MIEP, agreed that if a peace agreement were reached and international authorization provided, a contingent of 120 to 150 Hungarian troops should participate as KFOR peacekeepers.[54]

During the conflict Istvan Csurka (MIEP) talked about redrawing the map after the conflict so Vojvodina Hungarians could have independence. The government and HSP-AFD opposition promptly dismissed his remarks as irresponsible. In early May, though, when Zsolt Lanyi, deputy chairman of the coalition Smallholders' Party, publicly suggested the possibility of independence for Vojvodina, the government immediately disapproved the statement.[55] After Yugoslavia accepted the NATO peace plan (3 June), many of Hungary's internal tensions and pressures were relieved.

One of the lessons learned by the opposition HSP from the Kosovo action was the need to get a two-thirds parliamentary majority to authorize *any* future combat action from Hungarian territory. They submitted a draft constitutional amendment to this effect in the fall of 1999.[56] In early December Tamas Wachsler (Fidesz) and Ferenc Juhasz (HSP) forged agreement regarding military troop deployments and on constitutional amendments. The government would retain power over troop movements and military exercises, but the sending of Hungarian peacekeepers abroad and the launching of an attack against another country from Hungarian soil would require a two-thirds parliamentary vote.[57]

In sum, Hungary initially appeared quite restrained in its support of the NATO operation in Yugoslavia, making it clear that ground forces were out of the question. The opposition HSP attempted to reconsider the five-party agreement to provide NATO with air corridors if NATO intended to carry out air combat operations from Hungarian soil. Though blind-sighted during the war, the HSP could create future parliamentary constraints. While the prime minister seemed to suggest that NATO had "special" obligations to protect the Vojvodina Magyars, who were not citizens of Hungary, the United States and NATO did not share his perspective.

DEFENSE REFORM TENSIONS

Although Prime Minister Orban wanted to establish a super-ministry to formulate and coordinate policy, this proved difficult. This became apparent when three state secretaries, without permission, lobbied for a Lockheed official to become U.S. ambassador to Hungary. Chancery Minister Istvan Stumpf announced that two state secretaries—Gabriella Selmeczi and Istvan Balsay—had to resign. The third signatory—Bela Gyuricza, responsible for security and defense policy—ill with cancer, died. After only one year, the office of the prime minister, which had grown to 430 members, needed to be reorganized.[58] This was particularly an imperative in security and defense policy, since the defense ministry had been entrusted to Fidesz's Smallholder coalition partner Janos Szabo, who by his own admission, knew little about defense.

Prime Minister Viktor Orban also continued to raise public controversy on defense and security issues. In Canada in late October 1999, Orban created a political storm when he remarked on the possible deployment of nuclear weapons in Hungary.[59] A few weeks later, Orban again referred to the need for NATO to protect Hungarians in Vojvodina, raising another public controversy. In an effort to prevent another reoccurrence, the office of the prime minister added a communications secretariat headed by Istvan Toth to mold Hungary's image abroad with a budget of Ft4.3 billion for the year 2000.[60]

Defense State Secretary Tamas Wachsler noted publicly that the defense ministry was not involved in the Lockheed scandal and refused to comment about rumors of a critical prime minister's office assessment of the defense ministry. He did note that the General Staff and defense ministry merger would be completed in June 1999, and that many "feared for their jobs" because many departments and organizations would be unified or liquidated.[61]

Chief of General Staff Ferenc Vegh admitted to disagreements with the ministry over the planned integration, but felt they could be settled through negotiation, and that he had no plans to resign. Defense Minister Szabo stated that replacing the chief of General Staff was "not on the agenda."[62] In early June Tamas Wachsler added to the tension when he noted that the defense ministry's budget had to be readjusted due to the added cost of Hungarian troops in Kosovo. The political decision-makers could either grant additional funds to the ministry or tell the ministry which of its tasks need not be fulfilled.[63]

Civil-military friction increased as the General Staff's integration proceeded. Rumors persisted that one of the two deputy chiefs of staff, LTG Nandor Hollosi, would retire since under the reform the chief was to have only one deputy (LTG Lajos Urban). Hollosi, though, noted that he had signed a contract valid until December 2002 with Defense Minister Szabo. Chief of Staff Col. Gen. Ferenc Vegh added that Hollosi was an "excellent colleague" and that both his deputies had "serious tasks [and] their positions had been established by government decree."[64] When he later took over as chief of staff (August 1999), Lajos Fodor noted that in the course of the merger of the army staff and the defense ministry, the former military staff of one thousand people would be reduced to six hundred because the leadership of the army and the defense ministry were oversized, bureaucratic, and "too heavy at the top."[65]

Contributing to the civil-military tension was the fact that General Vegh wore two hats (see figure 5.2). As Chief of Staff Vegh was subordinate to Defense Minister Szabo, and as commander of the Hungarian army he had responsibility to report to President Arpad Goncz. In response to public rumors that Janos Szabo was upset that Vegh had supplied Goncz with the army staff's integration concept, Vegh responded that the president had "requested a written blueprint [and as] Commander of the Hungarian Army. . . . I had to inform [Goncz] of issues regarding the army on a monthly basis."[66]

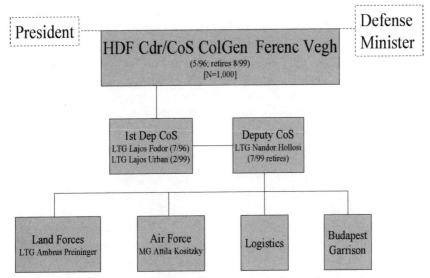

Figure 5.2.   Hungarian General Staff, 1998

Tamas Wachsler confirmed (1 July) that Janos Szabo had relieved LTG Nandor Hollosi of his post and placed him into retirement, and claimed that plans to remove Col. Gen. Ferenc Vegh were only rumors. Wachsler though admitted to "serious conflicts [adding] the minister wants to work with someone who supports him in carrying out the reform."[67] On behalf of the prime minister, Jeno Poda outlined three options for the chief of staff; either Vegh should accept and fulfill the government program, resign, or be sacked by the defense minister.[68] On 9 July Ferenc Vegh made up his mind and tendered his resignation to President Goncz, effective 1 August 1999.

The parliamentary HSP opposition issued a statement that "unjustified dismissals and personnel changes in the ministry and General Staff . . . cast doubts on the competence of the defense ministry leadership."[69] As if to confirm the HSP statement Defense Minister Szabo reversed his February 1999 decision to relieve Ambrus Preininger as land forces commander because Szabo now did "not consider Andras Havril suitable to manage the reorganization of the army."[70] (Air Force Commander Kositsky had already resigned in March.)

UNACHIEVABLE NATO FORCE GOALS AFTER KOSOVO

As the Kosovo air campaign began to wind down and the postconflict costs escalated, the defense ministry found its resources stretched. In mid-May

Hungarian Army Commander Ferenc Vegh noted that due to flood damage and the Kosovo crisis, the defense budget would not permit modernizing of MiG-29s or equipping helicopters with infrared guidance systems, adding that the equipping of aircraft with IFF systems and the equipping of regular troops alone would be a burden.[71] Two weeks later Vegh confirmed that the added Ft1.5 billion cost of the Kosovo conflict meant that the MiG-29 trainer that Defense Minister Szabo had promised (which cost Ft1 billion) would not be acquired in 1999.[72] In addition, the seventeen NATO-trained MiG-29 pilots, who were to fly one hundred hours, would now find their flight time reduced to eighty hours.[73] LTG Nandor Hollosi added that the MiG-29s, manufactured in 1993, needed to be examined and repaired by 2002, assuming that the Russians would deliver the spare components and driving mechanisms.[74]

Parliament approved a resolution (15 June) to offer a battalion to Kosovo to guard KFOR's main command post and telecommunications. The Hungarian contribution of a 320 to 350–man KFOR peacekeeping battalion to Kosovo was to cost some Ft4–5 billion a year. Since the finance ministry once again confirmed that it would *not* pay the bill, the defense ministry would have to shoulder the cost.[75] LTC Gyula Papp commanded the unit that was to be manned by professionals and contract soldiers who concluded six-month contracts, with possibility of a three-month extension at a monthly stipend of $1,000 to $1,500.[76] During preparation several soldiers had to be sent home for social problems, psychological reasons, and for drinking. On 15 July, 324 soldiers (among them 16 women) left for Kosovo.[77]

To prepare for KFOR unit rotation in January 2000, another three hundred–troop unit was established at Debrecen. One-half of the Hungarian KFOR troops were to be replaced every three months.[78] Hungary also maintained 235 troops in SFOR. This presence would be reduced to 205 engineering troops in Bosnia (Okucani, Croatia) and extended through the end of 2000 (at a cost of Ft2.2 billion in 1999 and Ft2.9 billion in 2000),[79] and another Hungarian 109 peacekeepers remained in Cyprus under United Nations Peacekeeping Force in Cyprus (UNFICYP).

Since the defense ministry had to curtail its budget by Ft10 billion, it could cancel some exercises, reduce training (to include not acquiring the MiG-29 trainer), call up fewer conscripts, or postpone some NATO force goals. Confirming that the KFOR costs for 1999 would be Ft4–5 billion, Defense Deputy State Secretary Janos Karasz announced the ministry would not be able to fulfill many of Hungary's NATO obligations for the current year.[80] Postponements would include developing information technology within the army, upgrading communications of Hungary's NATO–designated troops, rendering the Russian BTR armored vehicles and T-72s NATO compatible, purchasing missing elements of the MiG-29 IFF, and ensuring one hundred hours for the twelve NATO-designated MiG-29 pilots.[81]

Hungary's communications requirements would be difficult to meet because its telecommunications equipment had not been modernized since the 1970s, the Janos Bolyai Military Academy was only producing one-third the necessary personnel, and communications officers were difficult to retain because of private sector competition.[82] Defense Deputy State Secretary for Personnel Policy Colonel Janos Szabo noted that it cost Ft10–14 million for the four-year training of a young officer, that the army lacked a strategy to keep the officers within the military, and that retention was an important problem and because of differences of views "there had been no human resource planning and management in the Army."[83]

When U.S. Secretary of Defense William Cohen visited Hungary in mid-July 1999, he publicly acknowledged Hungary's support during the Kosovo conflict, but stressed that Hungary needed to keep its promise as a NATO member to increase its defense spending.[84] The United States had financed Hungary's Ft800 million airspace sovereignty program built by Siemens, allowing Hungary to join NATO's communication and information system. NATO made substantial contributions to Hungary's defense effort by financing Ft23 billion ($100 million) for radars. The tender launched in 2000 sought completion by 2004–2005. NATO also would finance a project for the modernization of Hungarian airports estimated to cost tens of billions of forints.[85]

STRATEGIC REVIEW

Three months after joining the Alliance Defense Ministry, Administrative State Secretary Tamas Wachsler asked NATO (24 June) to extend Hungary's end-of-June defense plan deadline because the "defense budget needed to be regrouped." NATO approved the extension for the end of September to allow Hungary to perform a new strategic review. Aside from the financial burdens, Wachsler admitted to problems in standing up the KFOR unit, noting "the Hungarian Army is not really prepared to meet such tasks because of the lack of technological equipment."[86]

On 14 July Viktor Orban announced the Cabinet had passed a secret resolution—a so-called strategic review—designed to "help create an Army that can meet the challenges of the future."[87] Two days later, Deputy State Secretary for Defense LTG Lajos Fodor admitted that the previous Gyula Horn government had promised too much to NATO, that Hungary lacked the funds to complete all its planned tasks and would need to revise its commitments, and that the defense ministry was starting a strategic review to determine the future size of Hungary's forces. Fodor confirmed that he supported maintaining only one wing (not two) of fighter aircraft and increasing the number of combat helicopters to eighty or one hundred within a few years.[88]

Tamas Wachsler led a ten-member committee to work out the details of the army reform in accordance with the government decision on strategic review. The 21 July government decree on strategic review ordered the defense minister by 30 September to propose a far smaller military structure that was suitable to carry out its tasks and could be financed in the long term. The decree covered the army budget until 2004, calling for a total of Ft192.3 billion in 2000 (1.51 percent of GDP). Though the 2000 budget figure failed to meet the government's earlier commitment to increase the defense budget by 0.1 percent, Hungary argued that the cost of the Border Guard, which was in the interior ministry, was not reflected in the defense budget. Also the 21 July decree noted that by 31 March 2000, Defense Minister Szabo would propose "which organizations not directly involved in defense (e.g., health institutions and artistic ensembles)" that he intended to remove from his jurisdiction. The decree also ruled that the State Privatization and Asset Management Corporation (APV) should take over redundant defense ministry real estate.[89]

It was also apparent that outside analytical studies by Cubic Applications and RAND were having an influence. The Cubic study concluded that because of past force reductions, the Hungarian army had too many weapons (to include T-55s and 100mm anti-tank guns), and too much wartime reserve. It noted that Hungarian pilots did not fly enough hours, that MiG-21s should be withdrawn from service, and that the land forces' logistics supply and technological equipment was outdated. On the personnel side, it also argued that the army had too many colonels and lieutenant colonels, that an appropriate merit system was needed to evaluate performance and promotion, and that an appropriate recruitment system for warrant officers and contract soldiers should be established. The RAND study gave priority to helicopters, and argued against purchasing new fighter aircraft.[90]

Zoltan Martinusz, who replaced LTG Lajos Fodor as defense ministry deputy state secretary, noted in reference to the strategic review that though Hungary needed a more flexible, mobile, and smaller army, no definite plans on personnel cuts had yet been made. He added that an unhealthy situation existed in that while a total of twenty to twenty-one thousand troops were serving as individual soldiers, twice as many soldiers were in support functions and in various commands and other military organizations.[91] In discussions about the appropriate size of Hungarian Defense Forces, Ferenc Gazdag of the Defense Ministry Institute for Strategic and Defense Studies noted that historically Hungary's armed forces had been limited twice in the twentieth century. Article 104 of the 1920 Trianon Treaty limited the defeated Hungarian army to 35,000 troops, while Article 12 of the 1947 Paris Peace Treaty limited it to 65,000 troops.[92] Defense Minister Janos Szabo noted that the 60,000 (52,200 troops and 7,800 civilians) in 1999 would likely decline to roughly 45,000.[93]

The 12,500-troop air force is to be reduced by three thousand after the deployment of the NATO radar system by 2005. Some two thousand troops were to be demobilized from this unit alone, because eight hundred soldiers will be able to accomplish this task with more advanced technology. Air Force Chief MG Istvan Talla noted that Hungary's helicopters would not be replaced, but rather modernized by the Hungarian Danube Aircraft Engineering Factory, and that he wanted pilots to fly new fighter aircraft by 2005.[94] Despite Talla's wish, *Nepszabadsag* announced that the Hungarian army would only have one aviation unit (MiG-29s at Kecskemet), that financial resources were not available for modernizing the MiG-21 regiment at Papa, which would expire in 2001 and 2002, and that it was unlikely that new aircraft would be purchased before 2008.[95]

When details of the strategic review came to light, it became clear that the defense ministry's planning was not functioning. The air force helicopter squadron at Szentkiralyszabadja would move to the Papa Airbase when the MiG-21s were withdrawn from service. During 2000 the air force would withdraw MiG-21s, sell the Yak-52 trainers that the Antall government had purchased, sell its Mi-2 helicopters, and withdraw the L-39 jet trainers from service. Similar to the Su-22s, which were eliminated one year after their major repair in 1997, the L-39s, and the Yak-52s had recently had major repairs in the Danube Aircraft Factory.[96] Scarce defense resources were wasted again.

In addition to financial constraints, it was clear that Kosovo lessons and NATO commitments were influencing the strategic review. Defense Minister Janos Szabo specifically noted that Kosovo highlighted that Hungary needed a "radical transformation of the current system" and the defense ministry had "inherited difficulties [from] the NATO obligations accepted by the previous government."[97] Tamas Wachsler added that "Kosovo proved that the army in its current form cannot operate in the future . . . [adding] although the defense budget is steadily rising, it is still unable to conform to the minimum expectations posed by NATO's defense planning system."[98]

The strategic review of the Hungarian army also included intelligence and its oversight. Some wanted to subordinate the Military Security Office and the Military Intelligence Office to Laszlo Kover, minister in charge of secret services; and to reduce Hungary's five services to three. This reduction would be consistent with the defense ministry's effort to implement substantial personnel cuts and to eliminate parallel offices. Presumably the two military secret services would be subordinated to the chief detection department of the Hungarian army supervised by the chief of staff, rather than remaining under the defense minister's direct supervision. According to one source, the two offices comprised a combined staff of 2,357 people, and its 1999 budget of Ft7 billion would increase to Ft8.3 billion in 2000.[99] At the same time, though, members of the Parliamentary Defense Committee apparently resisted the government's efforts to amalgamate the intelligence services apparently in the belief that separation of the three civilian and two military secret services should be maintained in order to check on each other.[100]

The strategic review projected that roughly sixteen thousand people would be dismissed over the next two to three years, which would likely result in *increased* personnel cost obligations. Since the dismissal of 6,700 men in 1995–1996 cost Ft5 billion, some estimates for dismissing 16,000 reached Ft18–20 billion.[101] After Prime Minister Viktor Orban briefed opposition party leaders on the strategic review concept, the government approved it on 26 October 1999. On 28 October Defense Minister Szabo and Chief of Staff Col. Gen. Lajos Fodor issued a jointly signed public letter, noting that the Hungarian army had now "exceeded the critical level [confirming that the] size, composition, and command system of the Army [were] not in harmony with its system of tasks and with the budget possibilities," and that the size of the army staff would decline to forty-five thousand.[102]

The government decree decided *not* to separate the posts of commander and chief of staff of the Hungarian army. The original proposal, which envisioned separating the dual-hat position and make the commander subordinate to the defense minister and the chief of staff to the defense administrative state secretary, had contributed to Ferenc Vegh's resignation.[103] The strategic review eliminated the position of commander of the Hungarian army. Henceforth, Col. Gen. Fodor Lajos was simply chief of staff and in this capacity chaired a committee to work out the tasks of dropping the defense sector to forty-five thousand in two years' time (see figure 5.3).[104]

Another disconnect involved the turning over of defense assets to the APV. First, the APV did not agree with the defense ministry's Ft12.8 billion valuation for its 250–300 military installations, airports, barracks, and sports facilities, which it needed to pay for previously purchased Mistral missiles and T-72 tanks.[105] Finally, the APV agreed on 14 December to pay Ft9.8 billion for the Mistral missiles acquired in 1998; the other Ft3.0 billion would

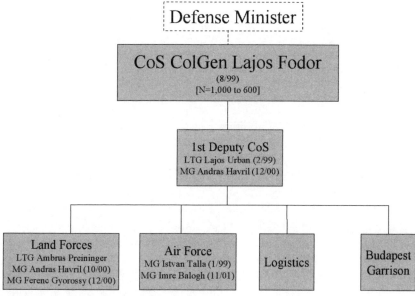

Figure 5.3.   Hungarian General Staff, 1999

be received later, but would be immediately returned to the APV to service the defense ministry's outstanding debt for the T-72 tanks acquired a few years earlier.[106]

Hungary finally presented its DPQ to NATO after a three-month delay in order to complete the strategic review. Zoltan Martiniusz claimed that NATO's assessment was now positive because "Hungarian Armed Forces reform plans are based on reality." NATO, though, did suggest that the defense ministry should "begin its planned technical developments sooner."[107]

The OSCE Istanbul Summit in November also bolstered Hungary's defense interests by committing Hungary to 10 percent reductions in CFE treaty limits. While the existing CFE treaty permits Hungary 835 MBTs (it only held 807), 1,700 APCs (it only held 1,332), 840 100mm artillery pieces (it only held 839), 180 combat aircraft (it only held 136), and 108 helicopters (it only held 59). If all the countries implemented reductions by 2002, Hungary's neighbors' weapons would drop significantly, while Hungary would only have to reduce by 97 MBTs (to 710) and 89 artillery pieces (to 750).[108]

Ferenc Juhasz (HSP), deputy chairman of the Parliament's Defense Committee, noted that the strategic review concept *must* succeed because NATO (and the EU) would consider Hungary "lacking in seriousness."[109] Juhasz expressed concern on the following issues. First, he claimed the strategic review was like an orchestra without notes or a director in that neither the goal nor the way to the goal was defined. The concept outlined further staff and troop reductions, rather than what the army wanted to have in 2010. Second, Juhasz claimed the internal budget was not transparent, specifically citing the impact of the Ft6–7 billion for SFOR and KFOR, the need to pay soldiers' insurance contributions in 2000, and to clarify how the National Health Insurance Fund would support the military after the transfer of military hospitals from the defense ministry. Juhasz estimated the "hidden" costs to be Ft30 billion. Third, he warned of the dangers of entrusting the army reform entirely to soldiers arguing that politicians should make decisions on the future of military service and new style military training.[110]

Defense Minister Szabo outlined the details of the strategic review concept. Noting that the detailed phasing would be presented to the government in February 2000, Szabo stressed the first priority of the ten-year program was to improve the living and work conditions of the military, to replace some military vehicles in 2000, and to renovate MiG-29s to attain NATO compatibility. Comprehensive technological improvements to the armed forces would occur only after 2005.[111]

When a number of soldiers died during an outbreak of meningitis in an army barracks in Kalocsa (Southeast Hungary) in December 1999, public attention focused on the state of neglect in the army and the need to refurbish barracks. In the strategic review, Defense Minister Szabo promised to improve soldiers' living conditions by providing a continuous, rather than periodic, hot water supply and maintaining a minimum temperature of 18–20 degrees Celsius.[112] The government had spent an average of only Ft300 million per year to modernize army barracks during the 1990s.[113] Now it would need to spend Ft25.2 billion between 2001 and 2003 to renew them and Ft9.4 billion to construct housing units.[114] The AFD's response to the meningitis outbreak was to call for abolishing compulsory service. Hungarian public opinion (57 percent) wanted to abolish compulsory service, while only 33 percent preferred to maintain it.[115]

After the government's National Security Cabinet approved the strategic reform on 16 March 2000, the Hungarian public debated its personnel, procurement, and basing ramifications. On personnel, although the regular army was to be reduced by five thousand by June 2001, the schedule was accelerated to January. The Army Trade Union proposed retraining those army officers leaving service and to help them find new jobs, possibly placing them with the police and the Border Guard who were struggling with major staff shortages.[116] In fact, the problem was less drastic than it appeared. While the plans called for eliminating 2,700 officer and 1,400 NCO positions, only 80 percent and 60 percent, respectively, were filled, and only 2,000 officers and 1,000 NCOs would need to be discharged.[117] In the end, the strategic reform's overall staff cut of 15,000 was less painful because 9,000 (60 percent) were conscripts who would not be called up, and 2,800 officers (almost 20 percent) would continue to work in nine public utility companies set up by the defense ministry. In addition, alternate employment possibilities for released defense personnel were available in penal institutions, Border Guard, Interior Ministry, and Finance Ministry.[118]

Deputy Defense State Secretary for Personnel Tividar Farkas reflected on Hungarian army personnel problems. By the end of 2000, Hungary had downsized its military personnel to 36,400 mostly by reducing the intake of conscripts. But restructuring the "inverted pyramid" in the officer corps would be a major challenge, because it remained unclear *how* to reconstruct the armed forces. Although there were roughly nine thousand deputy officers and eight thousand officers in the Hungarian army in the fall of 2000, the number of deputies should have been twice that of the officers. Added to this inverted pyramid problem was the fact that one thousand lower ranking officers expressed their intention of leaving the army rather than accept new positions after the reform.[119] By mid-2000 women also comprised 10 to 12 percent of the Hungarian army's officers. This proportion is higher than in the United States, and much higher than the 1 percent of the German officer corps, all serving in the health sector.[120]

Second, by the end of 2000 although the Hungarian army needed 7,000 contract soldiers, it only maintained roughly 3,500 who earned approximately Ft30,000 ($120) per month after taxes. Defense Deputy State Secretary for Personnel Colonel Janos Szabo explained that the ministry was exploring ways to increase contract soldiers' wages by 20 percent in 2001. But Defense State Secretary Tamas Wachsler noted that to meet the army's quota, the salary would have to be at least doubled. Since the budget's financial resources were insufficient, Wachsler believed "this staff level would not be reached."[121] Also a Hungarian Defense Ministry study suggested that 25 percent of the contract officers, who worked close to home, might quit as a result of the army reorganization.[122]

By the end of 2001 only approximately 4,000 of the 6,700 contract soldier positions were filled and more than 1,000 NCO positions remained unfilled. This meant that many units had an insufficient staff of contract soldiers that threatened their mobilization readiness. In the land forces roughly 30 percent of the military posts were filled and in the air force only 50 percent.[123] The NCO shortage also created problems; many NCOs had quit due to insecurity, lack of direction, and respect. Although four army recruitment offices with additional salary incentives had been established, they were still not sufficiently successful. By 2006 the army wanted to have thirteen thousand NCOs, nine thousand contract troops, and eight thousand officers.[124] To achieve this, the government proposed a new law for professional and contract soldiers that was to increase the number of NCOs and lower ranking officers by establishing a predictable career development path. As of 1 January 2002, the salaries of soldiers increased by 70 percent; the NCOs' pay would start at Ft104,000 per month, while starting officers would earn Ft130,000.[125] When the first phase of recruitment (September–November 2001) ended 2,400 NCO contracts had been signed.[126]

The conscription issue also festered. While Istvan Simicsko (Fidesz) claimed that conscription service would drop from nine to six months in 2002, within the ruling coalition the FKGP expressed skepticism. In the opposition, the AFD concluded a signature campaign list of 203,000 to abolish conscription.[127] Though the army conscript component was to be smaller, the shorter six-month training cycle meant that 30–35,000 conscripts would require training rather then 28,000. Added to the manpower problems was the unanimous decision of the Parliament's Defense Committee to commence with six-month conscription in January 2002 and to create volunteer reservists, who could be on alert for five years, undertake more frequent military service, and serve on missions abroad.[128] On 12 June 2001 parliament adopted a law that would reduce compulsory service to six months as of 1 January 2002 and mustered the necessary two-thirds majority to incorporate the Hungarian army into the defense ministry.

In August 2001, the HSP promised further systematic changes in defense if it won the 2002 elections. Gyorgy Keleti said that the HSP would end conscription, reinforce the NCO staff, and establish a professional army by the end of 2006. Former Deputy Chief of Staff Nandor Hollosi, now an HSP deputy, also criticized the Orban government claiming that Hungary still did *not* have a national security or military strategy.[129]

The most controversial issue of the reform—the merger of the military staff and the defense ministry—also meant that parallel operations would be eliminated. According to Col. Gen. Lajos Fodor only sixty-two of the current eighty-five general officer positions were filled in August 2000, and this was to be further reduced to forty-two.[130] Defense Minister Szabo confirmed that after the defense reform at least one hundred people would become superfluous from the six hundred-person general staff and three hundred-person defense ministry.

The strategic review projected slow procurement. During 1999 the purchase of the final twenty (of 487) BTR-80 armored personnel carriers was completed (for $240 million of Russian debt) and during 2000 the fifteen-year Ft200 billion program to modernize vehicles would begin. Defense

Minister Janos Szabo noted that procurement funding in 2000 would not exceed the Ft11 billion spent in 1999. Of this Ft4.5 billion was budgeted for modernization of vehicles, as well as renovation of one MiG-29 during 2000 and two simulators; one for the air force and one for the army.[131] But the procurement challenge was enormous! Because of the need to provide sanitary conditions in old barracks, Peter Dobo, Hungarian army logistics head, noted he would redirect Ft2.5 billion from the vehicle modernization program to barracks for 2000. Also in order to attract and retain soldiers, the army planned to construct and purchase 1,300 apartments by the end of 2003. But these plans were frustrated when widespread flooding again occurred in April 2000, necessitating another state of emergency and the need to employ sixteen thousand people (of which three thousand were soldiers) in the emergency effort. The emergency resulted in an unplanned Ft37.3 billion expenditure that was funded by deducting 2.1 percent from *every* government budget chapter to include Ft3.0 billion from defense.

Despite the budget squeeze, in July 2000 the Defense Ministry Procurement Office signed a statement of intent with DaimlerChrysler Aerospace (DASA) to modernize fourteen of Hungary's MiG-29s for Ft20 billion over three years.[132] U.S. Ambassador Peter Tufo criticized the decision, claiming that the aircraft would not be NATO compliant and added the United States would offer Hungary used F-16s in September. When Defense Minister Szabo noted that the letter of intent was not a binding contract,[133] a clamor of competitive bids arose. The Russians agreed to repurchase the Hungarian MiG-21s, which would be decommissioned when Papa Airbase closed at the end of August 2000, if Hungary would modernize their MiG-29s. Then the United States, Belgium, and Turkey offered to lease Hungary some of their surplus F-16s.

Zsolt Lanyi, Smallholder chairman of Parliament's Defense Committee, roundly criticized Defense Minister Szabo (also a Smallholder) for lack of information on the closure of Papa Airbase, the law on the unified system of authority in the defense ministry, and on the implementation of the army reform. On 31 August 2000, First Deputy Chief of Staff LTG Lajos Urban defended the Papa Airbase closing because the MiG-29s were stationed at Kecskemet Airbase and the army could only afford one combat aviation regiment.[134]

When Hungary decided to close the Papa Airbase, NATO accepted Taszar Airbase as an alternate for the Ft24 billion in NATO development funds allocated for three airbases (Ferihegy and Kecskemet). While the National Security Cabinet postponed the earlier defense ministry decision on modernizing the MiG-29s because DASA's repair costs were too high for a short-term solution, they did approve a minimal program that would provide

NATO-compatible radio and IFF equipment. The Cabinet also directed the defense ministry to conduct an expert study on the purchase of used Western combat aircraft.[135] Janos Szabo noted that the MiG-29 upgrade decision would be postponed until mid-January 2001, adding that a decision was needed since the fighters' life ended in 2003.[136] Confusion and uncertainty prevailed. Though the Papa Airbase was to be closed and the sixty pilots from the Papa Aviation Regiment were released, the Parliament Defense Committee adopted an amendment on 1 November 2000 to keep Papa open and to retain the Danube fleet.[137]

PRESIDENTIAL ELECTION AND
DEFENSE MINISTRY REFORM

Hungary's president is elected by a two-thirds majority of Parliament for a five-year term, with a two-term maximum. Arpad Goncz was elected in 1990 and reelected in 1995. His term expired in August 2000. According to the Constitution, if Parliament failed to achieve a two-thirds majority in the first two rounds, it can elect a president the third time with 50 percent plus one vote. In theory the ruling coalition, which had 54 percent of the seats, could elect its own candidate. But, both the coalition FKGP and opposition MIEP supported a direct election procedure. A further complication was the 1998 Fidesz-FKGP-MDF coalition agreement that gave the FKGP the right to nominate the presidential candidate, but neither Fidesz nor MDF would accept and vote for Jozsef Torgyan, chairman of the FKGP, for president. If the Smallholders' nomination was rejected, then it would threaten the survival of the coalition in its current form. (In fact during the Jozsef Antall government from 1990–1993, eight of the forty-four FKGP members, including Jozsef Torgyan, left the coalition and went into opposition.) In order to avert a political crisis, the parties came to a common solution and agreed to elect Ferenc Madl, an independent, when Goncz's term expired in August 2000.

In April 2000, the six parliamentary parties suspended negotiations on the integration of the defense ministry and army staff. After Ferenc Madl became president in August the talks resumed. On 12 October 2000, the participants agreed to eliminate the Hungarian army commander position and transfer its jurisdiction to the chief of staff; and they agreed to make the chief of staff and defense ministry administrative state secretary of *equal* rank and oblige them to cooperate with each other. The chief of staff also would have consultative rights on the Defense Council, which is summoned in crisis.[138]

Tamas Wachsler, who had pressed to have the chief of staff subordinate to him, resigned as administrative state secretary and was replaced by Tamas Perenyei, a former soldier, on 15 October.[139] Shortly after Col. Gen. Lajos Fodor and Defense Minister Janos Szabo met with recently elected President Ferenc Madl and agreed that the President's Military Office would provide liaison with the General Staff. While the Military Office would become smaller, President Madl wanted the chief of staff to provide him with monthly reports about the state of army reform.[140] In sum, Hungary managed to avoid a political crisis within the coalition by electing an independent president and eliminated a serious source of civil-military friction by eliminating the Hungarian army commander position and clarifying the relationship of the chief of staff to the defense ministry administrative state secretary and the president.

IMPATIENT NATO AND FRUSTRATED ARMY

NATO was becoming impatient with the pace of Hungary's defense reform. NATO SACEUR General Joseph Ralston, on a visit to Budapest in October 2000, said that he was "not dissatisfied, but impatient with the details of the Army reform." He argued that the pace of modernization should be faster, that the composition of the armed forces should be changed to comprise more NCOs than officers, and he called for the earliest closure of bases no longer required.[141] When Defense Minister Szabo met with NATO's Admiral Ellis in Naples, Italy, the AFSOUTH commander also expressed impatience with the pace of Hungarian reforms and urged the speedy filling of the NATO posts set aside for Hungary.[142] At the end of November, when a four-party delegation of the Parliament's Foreign Affairs Committee visited NATO headquarters, MDF delegate Miklos Csapody noted "our partners made it clear to us in very blunt language . . . we should accelerate the defense reforms."[143]

But frustration also was building within the army with implementation of the reform. In Hungary, military promotions and appointments are the prerogatives of the president of the Republic at the recommendation of the defense minister. But this time when Prime Minister Viktor Orban reviewed Defense Minister Szabo's final list of forty-two general officers, he erased ten names. Although Orban's criteria for exclusion remain unknown, they could not be explained on professional grounds. Later in talking with NATO Secretary General Robertson, Viktor Orban made no secret of the fact that he "personally participated in the shaping of the personnel decisions."[144] In the end, Defense Minister Janos Szabo appointed only thirty-two generals. Among the changes came the announcement that LTG Lajos Urban would be the land forces commander, and MG Andras Havril would be his deputy as of 1 December 2000.[145] But when LTG Lajos Urban was assigned to Naples, Janos Szabo announced that MG Ferenc Gyorossy would be the new land forces commander.

Throughout the armed forces the perception of promotions, based on political over professional criteria coupled with the slow pace of defense reform and modernization, reverberated on morale. Ferenc Base, director of the Defense Ministry Supply and Security Investment Office, noted that the amount for modernization would only grow significantly *after* 2006. Of the Ft23 billion spent in 2000, almost half went for clothing (Ft6.2 billion) and fuel (Ft4.1 billion). Of the Ft130 billion the Hungarian army would spend in 2001, Ft3.75 went for barracks renovation and construction, Ft4 billion to replace radios (twelve thousand at a cost of Ft10 billion over ten years), some buses, jeeps, and trucks (twelve thousand at a cost of Ft240 billion over fifteen years), Ft1.2 billion for the MiG-29 simulator, and the (still undetermined) cost of upgrading MiG-29s.[146]

NATO continued to criticize Hungary's annual increases of 0.1 percent of GDP in 1999 and 2000 defense expenditures (see table 4.3) because they were based on *planned*, as opposed to real, GDP growth. Hence, over the first two years of NATO membership when the economy outperformed its planned growth, the defense ministry *under*spent its target by almost Ft20 billion.[147] In mid-January 2001 Defense Minister Szabo announced he had decided not to buy MiG-29 simulators because the original Ft1.2 billion cost had risen to Ft4.0 billion and because the National Security Cabinet had still not made a final decision on MiG-29 upgrades.[148] Szabo (FKGP) also announced that although Smallholder leader Jozsef Torgyan had been forced to resign as minister of Agriculture and Rural Development over allegations of corruption, he would remain in his post as defense minister.[149]

Amid press reports that the government decided to acquire used F-16s, the National Security Cabinet decided on 8 February 2001 not to renovate the MiG-29s. Ferenc Juhasz of the HSP considered the decision "shocking" and asked the government to rescind its decision.[150] When the HSP and AFD delegates to the Parliament Defense Committee walked out in protest, Deputy Defense State Secretary Janos Karasz claimed that the Cabinet had made no decision and had only authorized the defense minister to prepare a decision on the basis of the offers.[151] In protest, the HSP also threatened to vote against the draft bill on integrating the defense ministry and General Staff and against the amendment to the constitution that would allow stationing or transiting of foreign troops in Hungary and participation of Hungarian troops abroad in war without prior parliamentary approval. Jeno Poda, the prime minister's national security advisor, took the threat seriously.[152]

The Hungarian government received two aircraft offers. As a counter to the U.S. offer of twenty-four F-16s, Saab-BAE offered twenty-four Gripens with promises of Ft160–170 billion offset investments in Hungary. Defense ministry experts prepared a comparative analysis of the two bids in early April. By the end of May the National Security Cabinet had still not made a decision. More offers from Belgium and Turkey and counteroffers emerged. The date for a decision kept getting postponed. At NATO in mid-June 2001 Viktor Orban noted that Hungary's resources were needed for health care, education, wages, and for boosting enterprises, therefore it was now mainly interested in leasing, rather than buying, aircraft.[153]

In the midst of the debate on fighters, Chief of Staff Lajos Fodor admitted that Hungary had "promised too many [pilot] flight hours to NATO."[154] Compared to NATO's annual norms of 160 to 180 hours, the Hungarian army leadership had promised 120 hours, but, in reality, they only flew an average of 60 to 80 hours. In addition, although 70 percent of the air force's MiGs were to be combat ready, less than 50 percent were because of a lack of spares and problems with electronic equipment.

In the midst of this, the National Security Cabinet added to the confusion when it once again reversed itself and decided in early April 2001 *not* to close down Papa Airbase and to maintain second-hand NATO compatible aircraft there. Ferenc Juhasz of the Parliamentary Defense Committee complained that the defense ministry under Smallholders' Party leadership was "scandalous and chaotic."[155] The decision also resulted in staffing shortages for several air force units. When the Papa Airbase closed after the withdrawal of MiG-21s in September 2000, seven hundred regiment troops transferred to other units. The reopening of Papa envisioned a staff of five hundred, many of whom would now return.[156]

During the summer the National Security Cabinet finally began evaluating the U.S. and Saab-BAE offers for fighters. Offers from Belgium and Turkey delayed the decision. Imre Mecs, former chairman of the Parliamentary Defense Committee, complained that there were no objective criteria to make a decision because Hungary had "no military strategy [and] . . . no strategy for the air force and air defense." Mecs therefore argued that procurement of aircraft was premature, adding that the MiG-29s should be repaired and extended.[157]

In yet another surprise move, the National Security Cabinet decided on 10 September 2001 to extend the life of fourteen MiG-29s until 2005 and to lease fourteen Gripens by the end of 2005 for twelve years. When Janos Szabo signed the MiG-29 agreement on 31 October, he noted the MiG-29 upgrade would cost roughly Ft5 billion (to include Ft2.2 billon for components) with work to commence in April 2002, permitting the aircraft to fly for three more years.[158] Defense Deputy State Secretary MG Janos Karasz noted the total cost of leasing, conversion, replacing components, training of crew, and buying armaments for the Gripen supersonic aircraft would cost Ft150 billion.[159] Although the opposition HSP agreed that the decision was economically sound, Ferenc Juhasz argued that the Ft150 billion was *not* available.[160] When Defense Minister Janos Szabo submitted the bill for leasing supersonic aircraft, it did not make reference to either the cost, partner, or type of aircraft, raising questions from the opposition HSP and AFD.[161] Despite this, Janos Szabo signed an agreement on 23 November to lease fourteen Gripens, with the first three to arrive in Hungary at the end of 2004, and the others by June 2005. In sum, the MiG-29 modernization would allow Hungary to meet its NATO obligation to provide eight MiGs through 2004, turning to Gripens in 2005.[162]

On 20 December 2001 the Hungarians agreed to lease fourteen Gripens (without weapons) for Ft108 billion, but including operation and training of staff. Hungary would make the first payment of Ft10 billion in 2002, and Ft20 billion in 2003. The estimated cost of weapons for the aircraft was Ft30 to 35 billion. The Swedes agreed to 110 percent offsets and would deliver three Gripens by the end of 2004. According to Economy Minister Georgy Matolcsy the contract would create nine thousand jobs in Hungary.[163] But a question that remained unanswered by the government was that by leasing only fourteen Gripens Hungary would *not* be able to meet its eight aircraft obligation to NATO.

After the decision had been made, Hungarian Air Force Commander MG Imre Balogh stipulated that roughly twenty pilots would be trained in Canada and Sweden to fly the Gripens, and that the Hungarian Gripens "will be equipped with a weapons capacity that exists in the airplanes of the Swedish Air Force, pursuant to the contract."[164] He reiterated the MiG-29s would fly through 2005, and that the Mi-24 helicopters needed modernization to include armaments, reconnaissance equipment to utilize the armaments, night-vision equipment, and NATO-compatible communication and identification equipment. In addition, the Mi-8 transport helicopters that had been indispensable for flood relief needed upgrading. Although the Visegrad Four agreed to modernize their 140 Mi-24 ground attack helicopters, Czech Defense Minister Tvrdik noted that Hungary presented "complications" in that "they do not have enough money or suitable legislation."[165]

As the end of the Orban-Janos Szabo mandate approached, problems with military personnel remained in evidence. The first 3,200 six-month conscripts began service on 12 March 2002. Of the total thirteen thousand conscripts in service, approximately eleven thousand were in units, one thousand were cadets, and the remaining conscripts were in the training base. Thus of the thirteen thousand conscripts, in reality only eleven thousand conscripts might be considered as potential combat power. Now that the length of conscript service has been reduced to six months, there is the additional problem of how to evaluate their effectiveness.

During 2000 and 2001 the 19,900 career cadres comprised 8,400 officers and 11,500 NCOs. In 2002 the 18,910 career cadres comprise 7,900 officers.[166] Active recruitment efforts and greatly increased salaries for contract soldiers drew interest, but limited results. During 2001 roughly three thousand applications for contract officers were filed. After the defense ministry announced increased wages (an average of 70 percent to Ft85,000 per month), another two thousand applied during December 2001 and January 2002.[167] Despite these public reports, many of the parliamentary authorized seven thousand contract officer posts still remained unfilled because most of the applicants were not qualified. According to BG Lajos Erdelyi, head of the Human Policy Department of the Hungarian army, during the 2001 campaign only five hundred candidates out of three thousand applicants actually fulfilled contracts. As a result, despite the increased wages and campaign efforts, Parliament authorization remained under quota leading to a hollow Hungarian army.[168]

At the end of his mandate, Defense Minister Janos Szabo on 7 March 2002 assessed his 2001 accomplishments. Szabo admitted that army reform had *only* really started in 2000 and that it would take until 2010 to complete. He claimed as accomplishments the merger of the General Staff with defense ministry in 2001, establishment of the legal basis to shorten conscript time to six months, and efforts to increase the proportion of contract personnel. Szabo noted the human resources measures taken to include reconstruction of barracks (Ft5 billion) and building and supplying of service flats (Ft3.1 billion) in 2001, with some 895 new apartments to be constructed in 2002. Szabo also referred to the decisions to lease fourteen Gripen aircraft, to prolong the service time of L-39 training aircraft, and the Visegrad agreement to modernize Mi-24 combat helicopters. He noted that the army staff received 8.75 percent salary increases and that civil servants' pay had increased by about 70 percent in two stages (in July 2001 and January 2002).[169]

At the same time HSP Prime Minister–designate Peter Medgyessy named Parliamentary Defense Committee Deputy Chairman Ferenc Juhasz as the next defense minister if the HSP won the upcoming elections. On 7 March 2002 Juhasz announced that if he were to become defense minister he would replace the conscript army with a voluntary, professional army by

2006, increase defense expenditures to 1.8 percent of GDP by 2004, and support the creation of a legal framework to guarantee a predictable career to professional soldiers. Juhasz also promised to develop a new national security and military strategy and emphasized that the politicization of the military that had occurred over the past three and a half years needed to end. He also promised to continue with barracks modernization and attempt to amend the constitution to allow Hungary to fulfill NATO Article 5 obligations.[170]

TERRORIST ATTACK, 11 SEPTEMBER 2001

When news of the attack against the World Trade Center and the Pentagon arrived, the prime minister, Fidesz-Hungarian Civic Party, and opposition HSP immediately condemned the acts and expressed support for the United States. MIEP Chairman Istvan Csurka expressed condolences, but noted that "U.S. policy is responsible . . . [because of] the immeasurable arrogance and expansion of globalism."[171] The AFD supported the other opposition parties and government, and declared Istvan Csurka's statement "repulsive."[172] The government offered to send doctors and rescue experts to the United States and strongly supported the 12 September NATO decision to invoke Article 5 for the first time.

On 24 September Parliament granted (272 for, and 12 MIEP against) the use of Hungary's airspace, designated airports, flight-control equipment, as well as command and logistical support, to the U.S. armed forces taking part in military operations.[173] On 25 September, though, when Parliament adopted a second resolution on Hungary's security and defense policy, the HSP abstained because it would not accept many amendments calling for a new national security policy, setting up a national antiterrorist council, and establishing a consultative body for detecting economic threats.[174]

Although the United States did not request direct military participation in the Afghanistan operation, it did send a list of eight items for NATO members' assistance. Following coordination of the six parties on 4 October, Foreign Minister Martonyi approved the U.S. request for NATO to use Hungarian airspace and airports in the antiterrorism operation, which an emergency parliamentary session approved on 5 October (286 in favor and 9 MIEP against).[175] In response to the U.S. request that if it had to withdraw units from the Balkans, the allies should replace those forces, Martonyi noted "for the time being, we are unable to propose anything, because we don't know what is needed and when."[176] For Hungary this was a sensitive issue because public opinion polls indicated that while 58 percent supported yielding Hungary's airspace, about 75 percent expressed concern about becoming involved in the war, and 94 percent were *opposed* to dispatching Hungarian soldiers to Afghanistan.[177]

Hungary did *not* feel threatened. On 7 October Lajos Fodor noted that since 11 September security measures had not been increased; some facilities were guarded, the readiness levels of land forces had not been raised, and only the air force was on alert.[178] After Ervin Demeter, chief of civilian secret services, together with the chiefs of the three other services reported to the Parliament's National Security Committee in a closed session, Gyorgy Keleti (HSP) noted that there were no threats against Hungary.

Janos Martonyi proposed a constitutional amendment that would allow the government to take measures without parliamentary authorization to apply Article 5 for mutual military assistance to NATO members.[179] Initially, the HSP had objections to expanding the government's role.[180] After NATO Secretary General Robertson complained to Peter Medgyessy (HSP) that Hungary's regulations were too complicated and requested they be simplified, the HSP supported the change.[181] On 17 November the HSP called for six-party negotiations to amend the constitution to permit foreign troops in Hungary when it becomes necessary and to agree on the acquisition of fighter aircraft.[182] Although the HSP accepted the proposal that the government could take decisions on NATO troop movements *in* Hungary, it rejected a constitutional amendment that would allow Hungary to send troops abroad, continuing to insist on a two-thirds Parliament vote.[183]

Martonyi later noted that European troops might have to replace U.S. troops in the Balkans. When asked if Hungary would also add troops to Macedonia, the foreign minister noted that "we are present in Bosnia and Kosovo and at the moment, this is what we are capable of."[184] Although the government maintained roughly five hundred peacekeepers in SFOR and KFOR, it asked Parliament for authorization to regroup troops within the existing 850-troop limit if necessary. This would permit Hungary to send thirty to forty peacekeepers to Macedonia if required.[185] (On 12 February 2002 the government approved sending forty peacekeepers to Macedonia.)

On 27 November 2001 Parliament adopted a package of legal regulations that authorized the government to introduce financial restrictions or bans on states, firms, and individuals suspected of terrorism.[186] Also on 14 December the government adopted a proposal to offer to send a medical unit of doctors and nurses to Afghanistan if needed.[187] The terrorist attack, coupled with six-month conscription that would commence in January 2002, also gave impetus to Istvan Simicsko's (Fidesz) proposal to create a National Guard. The government tasked the defense ministry to draft a legislative proposal that would require a two-thirds majority parliamentary vote to establish a National Guard that could deal with catastrophic situations, such as floods, and would provide basic military training to young people when conscription had ended.[188]

Another result of the terrorist attacks on the United States was the foreign ministry decision to establish a working group with the six parliamentary parties to draft the National Security Strategy by the end of 2001. Hungary would then write a National Military Strategy. During October the National Security Strategy draft underwent significant changes, employing more moderate tones, and no longer referred to neighbors, the Balkans, and Russia as "major risks." It did, though, raise from seventh to third priority the problems stemming from Hungarians living beyond Hungary's borders. Also the document no longer highlighted the need for an army of appropriate strength and capable of fulfilling relevant tasks, and made *no* reference to either the ten-year reform of the Hungarian armed forces or need to increase defense expenditures (since 1996).[189]

At the very end of the Fidesz mandate (6 May 2002), the Orban government adopted the National Security Strategy. But the HSP found the draft unacceptable and announced plans to amend it. The HSP criticized the National Security Strategy because it did not start from internal sources of danger, but from global challenges and included Ivan Simicsko's (Fidesz) concept of establishing a National Guard that was not supported by either the HSP or AFD.[190] The National Security Strategy, which had been a priority of the Orban mandate, came late and was virtually ensured to be short-lived.

In summary, the Orban government mandate dropped all efforts to adopt a new constitution, though a number of institutional and legal adaptations that were necessary, particularly with regards to NATO integration and obligations—such as deployment of Hungarian troops abroad—remained thorny problems. The prime minister's chancellery acquired greater powers particularly in the security and defense arena, and the National Security Cabinet assumed greater powers particularly in crisis management. Between 1990 and 1998 the Parliamentary Defense Committee chairman and the defense minister came from *different* parties of the governing coalition, constraining the committee's public critical review. Smallholder Zsolt Lanyi, who was of the same FKGP party as Defense Minister Janos Szabo, chaired the twenty-one-member Parliamentary Defense Committee between 1998–2002. Despite their common party ties, Lanyi often expressed frustration with Janos Szabo over the lack of information on the Papa Airbase closing, the system of authority within the unified defense ministry, and the implementation of the army reform.

The defense ministry under Janos Szabo evinced disarray and decline. Janos Szabo fired the opening salvo by dismissing Istvan Gyarmati for DPQ irregularities and continued to alter the staff by removing CoS General Ferenc Vegh, reducing the General Staff to one first deputy, dismissing the air force and land force commanders, Administrative State Secretary Tamas Wachsler, and Deputy State Secretaries Colonel Janos Szabo (Personnel) and Zoltan Martinusz (Defense Policy). Surprisingly, by the end of the Szabo mandate, similar to the 1994–1998 Keleti era, most of the ministry's state and deputy state secretary positions were filled by active or retired military officers. Although integration of the General Staff with the defense ministry had been a 1998 government program, it did not commence until the very end of the Orban mandate. The defense ministry's deficiencies were evident in the fields of defense planning (demonstrated by the need for a Strategic Review), budgeting (by the lack of internal transparency), personnel policy (by the meningitis outbreak and politicization of officer promotions), acquisitions and modernization (as exemplified by closure and reopening of the Papa Airbase and decommissioning of L-39s and Yak-52s after costly repairs).

Hungarian armed forces' personnel shortcomings were evident in that despite major pay increases and recruitment efforts, the four thousand contract officers that Szabo inherited in 1998 still remained at about four thousand at the end of the mandate in 2002. Although career soldiers were reduced by about 10 percent from 20,700 to 18,910, the "reverse officer pyramid" necessitated more substantial reductions at the upper levels of the officer corps affecting morale. Under Szabo, reductions in the armed forces were achieved mainly by reducing conscripts by 60 percent from 33,000 to 13,260, and by reducing their service time to six months, rendering them incapable of performing most military tasks. Despite NATO membership and integration obligations the government and defense ministry failed to fulfill its promises to increase defense expenditures and meet target force goals (such as maintaining adequate flying hours for NATO-designated pilots and by postponing many necessary acquisitions). Declining willingness to support NATO military operations was also evident during the 1999 Kosovo conflict. Because of Hungary's "special situation," Hungary refused to deploy its ground forces or allow any other NATO ground forces to participate in operations from its soil. Though after the conflict Hungary did provide 324 troops to Kosovo (KFOR) and further reduced its SFOR commitment to 205. Their lack of enthusiasm extended to Afghanistan after the 11 September 2001 terrorist attacks. Despite NATO's invoking Article 5, Hungary has *not* contributed ground forces to Operation Enduring Freedom (OEF) or International Security Assistance Force (ISAF) in Afghanistan.

6

2002 Parliamentary Elections: The Hungarian Socialists Return

The 2002 electoral campaign proved to be a heated one. In attempting to appeal to right-wing voters, Viktor Orban espoused nationalistic and anti-Semitic rhetoric, aggravating Hungary's domestic atmosphere and international relations with neighbors and the United States. Following in the wake of the Status Laws, calling for preferential treatment of Hungarians living in neighboring countries, on 27 January 2002 Orban called for linking the "economic living space (in Hungarian: *eletter*, with the Hitlerite connotation of 'lebensraum')" of Hungarians abroad with Hungarians in Hungary[1] to improve the country's economy. Graham Watson, chairman of the European parliamentary group noted these connotations were "incompatible with the conditions of admission to the EU as laid down by the Copenhagen Summit."[2] To distance himself from Orban, Peter Medgyessy, the HSP candidate for prime minister, campaigned on the platform of becoming prime minister of ten million Hungarians.[3]

On 20 February 2002 Viktor Orban reopened the historical question of the Benes Decrees, issued by Czechoslovakia in 1945, which stripped Germans and Hungarians of their citizenship. Orban claimed the decrees were at variance with EU law and needed to be revoked. Czech Prime Minister Milos Zeman, in response, refused to attend a meeting of the Visegrad Four prime ministers scheduled for 1 March 2002. The Hungarian Parliamentary Foreign Affairs Committee held an emergency session (27 February) to evaluate Orban's comments but failed to resolve party differences. Then on 4 March, the *Washington Post* carried a Jackson Diehl article titled "New NATO, Old Values," which expressed concerns about Hungary's political developments and stimulated debate in Hungary about the United States "meddling" in its electoral campaign.

In the end Viktor Orban's strategy of appealing to the right backfired, and the HSP returned to power in a striking upset. In a high turnout, with somewhat more than 70 percent of the electorate voting, the first round on 7 April determined 185 of Parliament's 386 seats; the HSP-AFD opposition won 98 seats, while the governing Fidesz-MDF won 87 seats.[4] On 10 April the AFD and HSP signed an electoral cooperation agreement, and the HSP withdrew its candidates from seven constituencies, while the AFD withdrew from the others. The second round on 21 April, which had a 73.5 percent turnout, resulted in the HSP holding 178 seats, the AFD holding 19, with one seat being a joint HSP-AFD, for a total of 198. Fidesz received 168 seats and the MDF 20, for an opposition total of 188.[5] In effect, one of the most significant results of the 2002 election was the reduction of the former six-party parliamentary coalition to four parties, since the FKGP and MIEP failed to make the 5 percent threshold.

Numerous judicial appeals capped the heated electoral campaign. When the Supreme Court rejected all complaints on 4 May, the National Election Committee declared the results official. President Ferenc Madl informed Peter Medgyessy that on 15 May he would charge him with forming a government, with the prime minister–designate claiming he would form a government by 27 May.

The new government (see figure 6.1) would have fourteen rather than seventeen ministries. The AFD would have four ministries—the Education, Information Technology and Communications, Environmental Protection

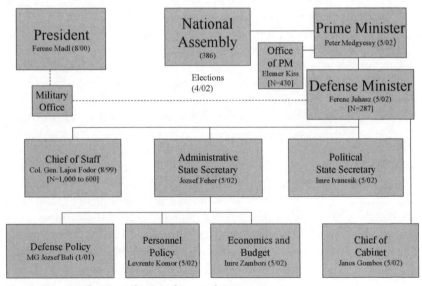

Figure 6.1.   Medgyessy/Juhasz Defense Reform, 2002

and Water Management, and Economic Affairs and Transport. In Parliament, twenty-five committees would operate with fourteen chaired by pro-government delegates and eleven by the opposition.

As Peter Medgyessy had earlier announced Ferenc Juhasz became defense minister. Juhasz appointed Imre Ivancsik, a member of the Parliamentary Defense Committee staff from 1998–2002, to be political state secretary and former Administrative State Secretary (1994–1998) Jozsef Feher returned to reassume the same position. Juhasz retained MG Jozsef Bali as deputy state secretary for defense policy, and appointed Levrente Komor, a retired military officer who had been teaching sociology, to head Personnel Policy, and Imre Zambori to head Economics and Budget. His chief of cabinet, Janos Gombos, was a longtime associate who assisted Juhasz in Parliament. Defense ministry division chiefs would also be replaced, while no immediate changes were to occur in the General Staff. Before Juhasz assumed his new position, President Madl on 16 May promoted BG Gyorgy Szaraz, who took a United Nations post, and Colonel Laszlo Toemboel to chief of staff of the air force.[6]

Defense Minister Juhasz, in an early interview, ruminated on Hungary's NATO challenges. Juhasz noted:

> During 2000, the year after Hungary joined NATO, we performed 76 percent of our commitments; the following year, the rate dropped to 50 percent, and last year it was only 26 percent. Essentially, we didn't meet any of our obligations; cooperation today is basically reduced to the Balkans mission; Hungary is increasingly regarded as a member failing to fulfill its commitments.[7]

In essence, the Medgyessy/Juhasz mandate would face the following major challenges: (1) to reorganize military personnel and meet the commitment to deliver an all-volunteer force (the AFD has promised to abolish conscription as soon as possible, the HSP by 2006); (2) "redraft a new national security strategy putting an emphasis on real threats [such as] international terrorism, international crime, migration-related risks, and potential ethnic disorder across the borders"; and (3) deal with a financially insolvent budget in light of the new long-term "burden" of leasing the fourteen Gripen fighter aircraft, which forced Hungary to withdraw its commitment to NATO because "these fighters will be limited only to the protection of Hungarian airspace."[8] Imre Mecs, an AFD MP, noted the government would respect its contract, but argued that it should negotiate with the Swedes to receive Gripens that would allow Hungary to fulfill its NATO common tasks.[9] On the day after taking office, Defense Minister Juhasz noted that he would ask for more funds to reach the NATO average of 2 percent by 2004 and begin personnel consultations and review the five-year contracts with senior civil servants and military recently signed by the outgoing defense minister.[10]

After two weeks in office, Juhasz confirmed there was a crisis in the defense ministry and noted:

> [W]e have found serious problems regarding the administrative, operational, and command structures. . . . we have very serious problems with development projects, increased wages, planned purchases, personnel activities, and integration. The situation does not fulfill the legal and other requirements of a democratic state's democratic army. . . . [adding] Unfortunately, the Army is in a bad mood and it resembles the army of a banana republic in many operational areas. I think that politics and politicians are to blame for this, people who recently did everything to politicize the Army.[11]

Over the next days, disclosure was made of efforts to launch new public procurement procedures in the wake of a mid-May 2002 decision to buy army short wave radios for Ft36 billion (as part of a ten-year Ft80 billion tender) and an attempt to renegotiate an extension of the 2002 installment of Ft24 billion for Gripen fighter aircraft. Defense State Secretary Imre Ivancsik claimed that the government would pay the Gripen installment from a special fund.[12] Defense Minister Juhasz noted a shortfall of Ft35 billion in the 2002 defense budget and claimed that he did not expect Hungary to be able to fulfill its NATO obligations.[13] On 22 June, Juhasz announced that he would create a second deputy post to the chief of staff to direct the staff and coordinate the work of the defense ministry's civilian and military organizations. As of 1 July, BG Zoltan Szenes would assume the position.[14]

In late July, Ferenc Juhasz requested patience and understanding when Hungary requested a delay in sending its Defense Planning Questionnaire (DPQ) to NATO from August to November 2002. He added that the current Ft264 billion defense budget was insufficient to meet army maintenance requirements, let alone for development.[15] On 30 July Imre Ivancsik began an investigation into the slow pace of development and shortcomings in the technical modernization of the army that would take six to eight months to complete.[16]

7

Civil-Military Relations: Prospects

Hungary's postcommunist political situation has remained relatively stable. All its elected governments since 1990 have completed their full four-year mandates: the MDF-led coalition in 1990; the HSP-led coalition in 1994; the Fidesz-led coalition in 1998; and the new HSP-led coalition, which began its mandate in 2002. The first and second rounds in all four postcommunist elections have been fairly consistent and impressive with turnouts of 65 and 45.5 percent, 69 and 55 percent, 56 and 57 percent, and 70 and 73.5 percent, respectively. Governments have changed without apparent turmoil. Despite the limitations of the October 1989 Constitution and the setback of the Miklos Nemeth defense reform in December 1989, Hungary has come a long way in its efforts to establish democratic oversight of the military.

Outlined below are the four main areas where change has taken place in Hungary:

1. *A clear division of authority between president and the government (prime minister and defense/interior minister) in constitutions or through public law.* The law should clearly establish who commands and controls the military and promotes military officers in peacetime, who holds emergency powers in crisis, and who has authority to make the transition to war. Underlining these formalities is evidence of the spirit of tolerance and respect for legitimacy between president and prime minister (government), who may often be from different political parties or ideological persuasions.

The existence of a constitutional and legal framework has resulted from Constitutional Court decisions that have effectively addressed the problems caused by the October 1989 Constitution and 1 December 1989 Defense Reform; the court's decisions have been respected and been incorporated in the 1993 National Defense Act and subsequent legislation. Despite a six-party consensus in 1994, continued wrangling over constitutional drafts between

1994 and 1998 rendered a new constitution into a dead issue after the 1998 elections.

Still Hungary needs to implement a number of other reforms to achieve effective civilian oversight of the military. Hungary's need to create an inter-agency organization that could formulate a national security policy (e.g., National Security Cabinet) began to take form under Viktor Orban's mandate. Under Prime Minister Orban, the National Security Cabinet brought together the ministers of Foreign Affairs, Defense, Interior, Finance, and the minister without portfolio to formulate policy and provide clear direction to the armed forces. While Prime Minister Viktor Orban attempted to use his enlarged chancery to perform this function, it did not succeed in part because of his erratic policies, politicization of the military, and because the defense ministry was in the hands of Smallholder coalition-partner Janos Szabo. In part this contributed to frequent reversals in defense policy to include Hungary's failure to meet its earlier agreed upon NATO force goal commitments and need to formulate a Strategic Review, as well as the reversals on issues such as the Papa Airbase, whether or not to refurbish MiG-29s, and the decision to lease Gripen fighter aircraft.

2. *Parliamentary oversight of the military through control of the defense budget and defense committee.* Parliament's role in deploying armed forces in peacetime, emergency, and war must be clear. Underlining these formalities is the need for the Defense and Security and Foreign Affairs Committees to provide minority and opposition parties with transparent information and allow consultation particularly on normal policy issues such as defense budgets, defense plans, and monitoring of the armed forces and on extraordinary commissions investigating defense/security violations. Parliamentary committees need staff expertise and adequate information to provide oversight and effective liaison with defense and interior ministries to help develop bipartisan consensus on defense and security matters. Similarly intelligence oversight committees should provide access to opposition parties.

Although parliamentary oversight has been one of the most problematic areas in achieving democratic oversight in postcommunist regimes, Hungary has done fairly well. In part this is the result of a fairly stable and predictable political process that has nurtured a number of parliamentarians who have developed a defense expertise. Indeed, following Hungary's first civilian defense minister Lajos Fur (1990–1994) were HSP parliamentarians retired Colonel Gyorgy Keleti (1994–1998), who had long experience with the military, and Ferenc Juhasz (2002 to present), who had been deputy chairman of the Defense Committee (1998–2002). Only Smallholder Janos Szabo (1998–2002), by his own admission, was ill-prepared for the defense post.

Since ruling coalition parties of the defense ministry traditionally have chaired parliamentary committees, the defense committee has exhibited greater restraint during major issues and blowups. For example this was evident when Defense Minister Keleti bought T-72s from Belarus and deployed MiG-29s to Poland for a PFP exercise without parliamentary approval. The opposition parties caused a stir, but the defense minister retained his position. During 1998–2002 Zsolt Lanyi, Smallholder chairman of the Defense Committee, often criticized his Smallholder Defense Minister Janos Szabo for poor management and the slow pace of developing the strategic review.

Two-thirds parliamentary majority voting requirements, which had been a problem for permitting the use of Hungarian airspace and airbases for NATO operations, have been resolved in the government's favor. But because the HSP believed that it was "blind-sighted" during the 1999 Kosovo bombing campaign, parliamentary decisions to launch bombing raids from Hungarian soil and to deploy Hungarian soldiers abroad even to fulfill a NATO obligation still require a parliamentary two-thirds majority of the delegates present. Though the new HSP government wants to change this, it remains to be seen if this will happen.

3. *Peacetime government oversight of General Staffs and military commanders through civilian defense ministries.* Defense ministry management should include preparation of the defense budget, access to intelligence, strategic planning, and defense planning to include force structure development and deployment, arms acquisitions and property management, and personnel policies to include career development and military promotions. The defense ministry also needs press and legislative liaison capabilities to communicate to the public and parliament how and why public resources are being expended.

In order to accomplish these objectives, defense ministries need "real" civilian defense ministers. Two common pitfalls need to be avoided: First, defense ministries can be politicized at times because the defense minister and/or state secretaries or deputy defense ministers are from different (majority or minority coalition) political parties (e.g., during the Orban mandate Janos Szabo and Administrative State Secretary Tamas Wachsler). Second, if retired military officers at times become "civilian" defense ministers, insider peer pressure could be hard to overcome to provide *real* defense ministry oversight (e.g., Gyorgy Keleti during the Horn mandate).

In the early 1990s many problems resulted, in part, from the scarcity of legitimate defense experts—whether civilian or military—who are capable of making the defense and security case to their legislatures and broader public though legislative liaison and public affairs directorates. In sum, the

Hungarian defense ministry needs responsible and capable civilian personnel to fulfill these functions to ensure that the ministry maintains real civilian oversight of the military. Twelve years after the 1990 transition, problems and shortcomings in the Hungarian defense ministry still remain evident and manifold. Though formal integration of the ministry and General Staff has finally begun at the end of 2001, little progress has been made yet in developing a viable defense ministry with oversight of the General Staff.

First, the defense minister is required by parliamentary regulations to report to Parliament every year on the defense policy and state of the Hungarian Defense Forces. Though the defense ministry prepares such a report on a confidential basis, the defense ministry's deficiencies were publicly confirmed by its need to revise its NATO Force Goals negotiated after receiving an invitation to join the Alliance in July 1997 and by the lengthy period it required to perform a Strategic Review *after* getting into NATO. This was confirmed again in 2002 when Defense Minister Juhasz requested NATO's patience for yet another delay in Hungary's force review. Neither Defense Ministers Keleti nor Szabo ever presented annual assessments to the Hungarian Parliament or society to give a public reckoning of Hungary's failings with NATO nor explained why or how these could be rectified.

Second, the General Staff's continued separation from the defense ministry until 2001 has also contributed to Hungary's problems. The General Staff, generally two to three times the size of the defense ministry, maintained its own chain of command to the defense minister; hence, there was a duplication of functions and no formalized means for cross-fertilization between action officers in the General Staff and defense ministry. Hence, very few horizontal contacts and communications between the working levels of the defense ministry and General Staff existed until recently. The fact that most of the positions in the defense ministry are filled by active duty or retired military officers does not improve communication with the General Staff because military officers tend to remain in the defense ministry for long periods, often with no idea of when, or if they will rotate to other military positions. The fact that many military officers who have retired remain in defense ministry positions as civilians tends to widen the divide. This situation can be improved only if military officers routinely rotate into the defense ministry and back to the General Staff. Also it remains to be seen if the decisions to eliminate the commander of the Hungarian army position and retain only the post of chief of staff in 1998 and to unify the General Staff with the defense ministry in 2001 will correct some of the institutional deficiencies that have remained evident for over a decade.

Third, the Hungarian defense ministry has been struggling to implement a mid- and long-term planning mechanism, which has been under development for years. This problem had earlier been highlighted by Hungary's participation in PFP, in frequent restructuring of the Hungarian Defense Forces, and in the need to participate and develop a Planning and Review Program (PARP). This planning mechanism failure was evident *after* Hungary became a NATO member in developing its Defense Planning Questionnaire (DPQ) and in the need to revise its Force Goals. It was also again evident in Juhasz's 2002 request to delay force goal negotiations with NATO. Though the defense ministry had made several attempts to develop a modified Planning, Programming, and Budgeting System (PPBS) system and the Defense Resource Planning Group developed a Defense Resource Management Model for Hungary, difficulty has resulted because in Hungary resource allocations run from top of the hierarchy down, rather than from bottom up.

Though procurement has been limited because of severe fiscal constraints, acquisitions problems remain, in part, due to unpredictable and fluctuating budget allocations, personnel incompetence, and political interference. Defense budget constraints have been aggravated by unfunded acquisitions that create future budget time bombs, such as the decisions in 1995 for radars, in 1997 for Mistral-2 air defense missiles, and in 2001 for fighter aircraft. Defense budget fluctuations resulted in 1999 when the finance ministry refused to shoulder the Ft4–5 billion for KFOR deployment and in 2000, when floods necessitated a 2.1 percent defence budget reduction (Ft3.0 billion), and when the actual economy proved better than planned in 1999 and 2000. As a result, the defense budget was pegged to the *lower* planned (as opposed to the actual) GDP figure (resulting in a shortfall of Ft20 billion). Gyorgy Keleti's decision to buy T-72s from Belarus had little to do with defense ministry planning or advice from the General Staff. Even after costly decisions for repair were made, as with Su-22s, L-39s, and Yak-52s, the aircraft were soon withdrawn from service. Finally, Viktor Orban's decision to lease fourteen Gripens in September 2001 was personal, made contrary to the defense ministry and General Staff proposal, and despite government promises of "extra-budget" funding, remains an unfunded acquisition that, like the L-159 aircraft in the Czech Republic, is a slow-ticking time bomb!

Hungary's need to restructure the defense ministry into an integrated institution, linking the defense minister (and his administrative and policy advisors) directly to the command structure, has only begun in 2002 and was reflected in the appointment of BG Zoltan Szenes. If successfully implemented, integration might facilitate the flow of defense needs from the armed forces to the government, opening up defense policies and activities to public scrutiny and accountability. In sum, when finally implemented, the integrated defense ministry should be more efficient and provide more effective oversight of the armed forces. But this necessary structural change is a decade-long task.

The defense ministry continues to shoulder the burden of having become a "retirement home" for military officers. The problem of "retired" or active duty military in key defense ministry positions, evident under Gyorgy Keleti, was just as pronounced under Janos Szabo with all key positions occupied by military officers—Tamas Perenyei, MG Jozsef Bali, MG Janos Karasz, and Col. Janos Szabo. The ministry has still not yet cultivated civilian specialists, nor has it developed a defense constituency in parliament or in Hungarian society. If Hungary is to develop the necessary political-military planning processes in order to effectively integrate with NATO, these difficulties will need to be overcome.

4. *Restoration of military prestige, trustworthiness, and accountability for the armed forces to be effective.* Because the military was controlled by the Soviet High Command through the Warsaw Pact (and top-secret statute system) and often used as an instrument of external or internal oppression during communist rule, postcommunist General Staffs must be apolitical and professional, and society needs to perceive the military as being socially useful, militarily capable, and under effective civilian control. In addition to the necessary constitutional and institutional arrangements, there needs to be a Legal Framework and Code of Conduct for professional soldiers and conscript citizens that would allow them to disobey illegal orders.

Military training levels and equipment must also be sufficient to allow a necessary level of professional prestige. This requires social support for conscription (for the foreseeable future) and a predictable stream of material resources (adequate defense budgets) that the defense ministry can "sell" to parliament and the broader society. In 2002, the Central European reality was different. Most militaries retained 25–35 percent of their 1988 manpower levels and 25–35 percent of the defense budgets in real terms. In sum, their readiness, training, and modernization levels have deteriorated significantly; in some cases, even to raise questions about their capacity to defend their territory let alone to participate in coalition defense tasks. Finally, though the Central European militaries have evidenced significant reforms and been restructured to accommodate NATO, force modernization continues to be greatly restrained by scarce resources.

Among the Central European militaries in NATO, Hungary has the most severe problems with its armed forces. The continued inability to solve the "reverse officer pyramid" problem with its professional officers, to build an NCO corps, to provide an adequate base of contract soldiers,

and to make conscription palatable remains evident. The relationship between Hungarian armed forces and society is dysfunctional; Hungarian
politicians and society still see the military as a costly luxury without much
social utility.

Conscription service has provided the main contact between society
and the armed forces, and that experience has been very significant in *erod-
ing* social support for defense. Poor physical conditions coupled with the
lack of meaningful training—as the December 1999 meningitis outbreak
confirmed—have contributed to the society's perception that conscription is
a waste of time and should be avoided. Hungarian politicians of *all* political
parties have derived popular support by reducing the intake of conscripts
and shortening service time; the MDF reduced the Hungarian armed forces'
81,000 eighteen-month conscripts to 51,560 twelve-month conscripts in
1994; the HSP to 33,000 nine-month conscripts in 1998; and Fidesz to
13,260 six-month conscripts in 2002. The new HSP government plans to
abolish conscription totally and to create an all-volunteer force by 2006!

Because the Hungarian public does not hold the armed forces in high
esteem, a vicious circle prevails. Government efforts since 1997 to develop
an adequate number of contract soldiers to fill the void created first by nine-
month and then six-month conscripts have consistently failed. Despite recent active public recruitment efforts and substantial pay increases, only
four thousand of parliament's authorized quota of seven thousand positions
have been filled; a number that has remained virtually unchanged since
1998. This shortfall seriously reduces the armed forces' capabilities and
raises questions about Hungary's ability to meet its goal of having nine thousand contract soldiers by 2006.

The NCO corps also presents a challenge. Efforts to build an NCO
corps have proven to be difficult. In 1992, the 22,900 career soldiers in the
armed forces included 8,500 NCOs and 14,400 officers. By 1997, the
21,346 career soldiers witnessed an increase to 10,120 NCOs and decrease
to 11,295 officers. By 2002, the 18,900 career soldiers saw a further increase
of NCOs to 11,500 and decrease in officers to 7,900. Hungary's plan is to
build up to thirteen thousand NCOs in the armed forces and maintain eight
thousand officers by 2006. But here the "inverted officer corps pyramid"
would frustrate the effort. While an *ideal* pyramid would require a ratio of
twice as many junior officers to officers, in early 2001 the two groups were
about equal. Compounding this problem is the fact that many junior officers—who also are now comprised of 10–12 percent women—have indicated their intention of leaving the armed forces.

To date, Hungarian armed forces "reform" has predominantly focused
on downsizing for political gain, rather than real reform. For example, the

Jozsef Antall MDF mandate of 1990–1994 did reduce the career element by 27.5 percent from 29,700 to 22,100, but to garner public opinion substantially reduced the intake of conscripts by 44 percent from 81,000 to 51,560. The Gyula Horn HSP mandate of 1994–1998 witnessed only a very slight further reduction of the career forces by 6.3 percent to 20,700 soldiers, but continued substantial reduction of conscripts by 40 percent to 33,000. The Viktor Orban Fidesz mandate of 1998–2002 continued the same policy; only a slight reduction of career soldiers by 8.5 percent to 18,910, and substantial reduction of conscripts by 60 percent to 13,260. Hence, the new Peter Medgyessy HSP government has no choice but to face the serious challenge of implementing *real* military reform and restructuring. If it doesn't, Hungary will not have a viable military.

Adding to this challenge, Hungarian societal support for military and defense operations appears to be on the decline. Hungary's 1996 commitment of five hundred engineering troops to support peace support operations in Bosnia in the IFOR was limited by finances to 416, and then in 1997 in the SFOR to 310. Although support was evident in the February 1998 Iraq campaign by offering airspace, airports, and medical doctors if an operation became necessary, *after* becoming an Alliance member Hungary's support for such operations has declined. Restraint was certainly evident during NATO's 1999 Kosovo bombing campaign where Hungary was hesitant to employ ground forces because of Hungarians in Vojvodina, and despite NATO's invoking Article 5, its support for Operation Enduring Freedom and International Security Assistance Force in 2001–2002 was weakened by refusing to consider the deployment of Hungarian soldiers to Afghanistan and offering only medical doctors.

Military training and force modernization needs significant attention and development to meet NATO standards. Hungarian defense expenditures of 1.4–1.6 percent of GDP remain the lowest of all its Central European neighbors. It has been simply inadequate and reflects the fact that respective Hungarian governments and their defense ministries have been either unwilling and/or unable to develop sufficient public understanding of the obligations and costs of NATO integration and have failed to develop parliamentary support for adequate defense budget levels.

Notes

CHAPTER 1

1. See Felipe Aguero, *Soldiers, Civilians and Democracy: Post-Franco Spain in Comparative Perspective* (Baltimore: Johns Hopkins University Press, 1995).

2. Janos Kis, the founder and leader of the Alliance of Free Democrats, made this observation arguing that the new Government must demonstrate economic success in order to be accepted. See Janos Kis, "Postcommunist Politics In Hungary," *Journal of Democracy,* vol. 2, no. 3 (Summer 1991): 4–5.

3. Zoltan Pecze, *Civil Military Relations in Hungary, 1989–1996* (Groningen: Center for European Studies, Harmonie Paper 2, 1998), 17.

4. These factors were initially developed in my *NATO Enlargement and Central Europe: A Study in Civil-Military Relations* (Washington, D.C.: National Defense University Press, 1996), passim.

CHAPTER 2

1. Radio Budapest, 1 July 1989, cited in Zoltan D. Barany, "East European Forces in Transitions and Beyond," *East European Quarterly,* vol. XXVI, no. 1 (March 1992): 18.

2. *Heti Vilaggazdasag* (28 October 1989): 4–5. *FBIS-EER-89-138* (9 December 1989): 6. Also see Section 29 (2) and Article 29/A (1), *The Constitution of the Republic of Hungary,* p. 5.

3. Keith W. Crane, Steven W. Popper, and Barbara A. Kliszewski, *Civil-Military Relations in a Multiparty Democracy: Report of a RAND Conference,* R-3941-RC (August 1990): 25.

4. Aside from the president, the Defense Council's membership includes the speaker of the National Assembly, the leaders of the parties in the National

Assembly, the prime minister, the ministers, as well as the commander of the Hungarian Army and the chief-of-staff.

5. Article 19, *The Constitution of the Republic of Hungary*, pp. 2–3.

6. Article 35 (1)h, *The Constitution of the Republic of Hungary*, p. 8.

7. Zoltan D. Barany, "Major Reorganization of Hungary's Military Establishment," *RFER* (7 December 1989).

8. MG Jozsef Biro interview, Zagreb *DANAS* (9 January 1990): 58–59. *JPRS-EER-90-045* (4 April 1990): 8.

9. Budapest Domestic Service (1 December 1989). *FBIS-EEU-89-231* (4 December 1989): 57–58. On 1 March 1990, further amendments to this process were discussed and supported by Ferenc Karpati in the National Assembly. Budapest Domestic Service (1 March 1990). *FBIS-EEU-90-042* (2 March 1990): 50–51.

10. Ferenc Karpati, Budapest MTI (22 March 1990). *FBIS-EEU-90-057* (23 March 1990): 48.

11. LTG Laszlo Borsits, statement to Vienna Military Doctrine Seminar (19 January 1990), 8 (emphasis in original).

12. National Assembly session (18 October 1989), Budapest Domestic Service. *FBIS-EEU-89-201* (19 October 1989): 33.

13. Budapest MTI (3 January 1990). *FBIS-EEU-90-006* (9 January 1990): 47.

14. Parliament Defense Committee session amending Defense Law (20 February 1990). *FBIS-EEU-90-036* (22 February 1990): 60.

15. Budapest Domestic Service (2 March 1990). *FBIS-EEU-90-043* (5 March 1990): 41.

CHAPTER 3

1. For discussion of the election and party positions, see Barnabas Racz, "Political Pluralism in Hungary: the 1990 Elections," *Soviet Studies,* vol. 43, no. 1 (1991): 107–36; John R. Hibbing and Samuel C. Patterson, "A Democratic Legislature in the Making: The Historic Hungarian Elections of 1990," *Comparative Political Studies,* vol. 24, no. 4 (January 1992): 430–54.

2. Article 40/A (1), *The Constitution of the Republic of Hungary*, p. 9. Also see *Magyar Kozlony*, No. 59, 25 June 1990, pp. 1269–79. *JPRS-EER-90-125* (31 August 1990), especially p. 14.

3. Gabor Juhasz, "Proposed Rules for Secret Services," *Heti Vilaggazdasag*, 8 February 1992, pp. 76–78. *JPRS-EER-92-031* (13 March 1992): 21–22.

4. Interview with Col. Karoly Gyaraki, chief of Military Security Office, *Magyar Honved*, 19 October 1990. *JPRS-EER-90-161* (7 December 1990): 38.

5. Goncz had been interim president since 2 May 1990.

6. Personal interview, 12 April 1991; and *Tallozo*, No. 35, 31 August 1990, pp. 1673–1674. *JPRS-EER-90-137* (3 October 1990): 24–25. According to Defense

State Secretary Erno Raffey the exact authority of the office is "nowhere . . . defined . . . a legal gap [adding] authority should be defined by Goncz." *Tallozo*, No. 33 (17 August 1990). Ibid., p. 26.

7. Hungary had a civilian defense minister before the Communist takeover; Peter Veres (National Peasant Party) was defense minister from 1946–1947.

8. Lajos Fur interview, *Nepszabadsag*, 10 September 1990, pp. 1, 5. *FBIS-EEU-90-178* (13 September 1990): 36.

9. *Review of Parliamentary Oversight of the Hungarian MOD and Democratic Control of the Hungarian Defense Forces*, Study No. 810 (London: Ministry of Defence, Directorate of Management and Consultancy Services, February 1996), p. 32.

10. Lajos Fur interview, *Tallozo*, No. 4 (25 January 1991), pp. 162–63. *JPRS-EER-91-039* (28 March 1991): 4.

11. Ferenc Mernyo, "Army Crisis: Self-Centered," *Heti Vilaggazdasag*, 13 April 1991, pp. 77–79. *JPRS-EER-91-083* (14 June 1991): 31.

12. Budapest Domestic Service, 24 March 1991. *FBIS-EEU-91-058* (26 March 1991): 35.

13. Budapest Domestic Service, 2 April 1991. *FBIS-EEU-91-064* (3 April 1991): 22.

14. Bela Kiraly, "Military Reform: What Should Be Done?" *Nepszabadsag*, 13 April 1991, p. 6. *JPRS-EER-083* (14 June 1991): 33–34.

15. Budapest MTI, 19 August 1991. *FBIS-EEU-91-161* (20 August 1991): 21.

16. Budapest Kossuth Radio, 19 August 1991. *FBIS-EEU-91-161* (20 August 1991): 21.

17. MG Janos Deak interview, Budapest Kossuth Radio, 18 September 1991. *FBIS-EEU-91-182* (19 September 1991): 20.

18. *Magyar Kozlony*, No. 103 (26 September 1991), pp. 2,111–13,125. *JPRS-EER-01-168* (13 November 1991): 14–16.

19. *Nepszava*, 3 December 1991, p. 5. *FBIS-EEU-91-234* (5 December 1991): 17.

20. Article 19/B.(2), The Constitution of the Republic of Hungary, p. 3.

21. Budapest MTI, 25 May 1992. *FBIS-EEU-92-103* (28 May 1992): 12.

22. "National Defense Draft Bill," *Heti Vilaggazdasag*, 15 August 1992, p. 15. *JPRS-EER-92-128* (14 September 1992): 39.

23. Budapest Kossuth Radio, 14 September 1992. *FBIS-EEU-92-180* (16 September 1992): 11.

24. The Constitutional Court held that these powers, which were not assigned explicitly either to Parliament nor the president, should be assigned by default to the government. See *East European Constitutional Review*, vol. 3, no. 1 (Winter 1994), pp. 11–12.

25. Amendment to the Constitution of the Hungarian Republic passed by the National Assembly on 7 December 1993. *Magyar Kozlony*, No. 186, 24 December 1993, pp. 11, 129–30. *JPRS-EER-94-009* (16 March 1994): 87.

26. Budapest MTI, 7 December 1993. *FBIS-EEU-93-236* (10 December 1993): 26.

27. Budapest *Uj Magyarorszag*, 3 December 1992, p. 5. *JPRS-EER-93-001* (5 January 1993): 14.

28. Lajos Fur interview, *Bulgarska Armiya*, 15 March 1993, pp. 1, 4. *FBIS-EEU-93-052* (19 March 1993): 19.

29. Budapest MTI, 17 March 1993. *FBIS-EEU-93-053* (22 March 1993): 34.

30. Lajos Fur statement. Budapest MTI, 26 November 1993. *FBIS-EEU-93-228* (30 November 1993): 30.

31. *RFE/RL Daily Report*, No. 67 (7 April 1993), p. 4. By March 1994, Germany had supplied 150 million DM worth of military supply parts. On 9 March Germany agreed to provide supply parts for Hungary's helicopters free of charge. *RFE/RL Daily Report*, No. 48 (10 March 1994), p. 5.

32. Budapest Kossuth Radio Network, 2 June 1993. *FBIS-EEU-93-105* (3 June 1993): 15.

33. Defense Ministry Spokesman Colonel Lajos Erdelyi. Budapest *Nepszabadzag*, 27 December 1993, p. 4. *FBIS-EEU-93-002* (2 January 1994): 9.

34. Budapest MTI, 16 February 1993. *FBIS-EEU-93-030* (17 February 1993): 18–19.

35. *RFE/RL Daily Report*, No. 49 (12 March 1993): 5.

36. Budapest MTI, 14 April 1993. *FBIS-EEU-93-074* (20 April 1993): 22. For text, see *Magyar Kozlony*, No. 48 (23 April 1993), pp. 2701–05. *JPRS-EER-93-061* (1 July 1993): 28–32.

37. Budapest Kossuth Radio Network, 7 December 1993. *FBIS-EEU-93-234* (8 December 1993): 23; and Laszlo Szendrei interview, *Uj Magyarorszag*, 9 December 1993, p. 4. *FBIS-EEU-93-239* (15 December 1993): 25–26.

38. Budapest Kossuth Radio Network, 4 October 1993. *FBIS-EEU-93-191* (5 October 1993): 22.

39. Arpad Goncz interview, Amsterdam *De Volkskrant*, 5 October 1993, p. 4.

40. Budapest *Magyar Honved*, 8 October 1993, pp. 26–28. *JPRS-EER-93-128* (10 November 1993): 30–31.

41. Budapest MTI, 14 January 1994. *FBIS-EEU-94-010* (14 January 1994): 9.

42. Lajos Fur interview, *Pesti Hirlap*, 15 January 1994, p. 3. *FBIS-EEU-94-015* (24 January 1994): 24.

CHAPTER 4

1. *Hungarian Review*, Newsletter of the Embassy of Hungary (9 June 1994): 2.

2. In 1990, the Hungarian Democratic Forum (HDF) gained 42.5 percent of the parliamentary seats on the basis of 24.7 percent of the vote. For a discussion of Hungary's reelection law, see Andrew Arato, "Elections, Coalitions and Constitutionalism in Hungary," *East European Constitutional Review*, vol. 3, nos. 3 and 4 (Summer/Fall 1994): 26–32.

3. *RFE/RL Daily Report*, no. 188 (4 October 1994): 4–5; and Budapest MTI (5 October 1994). *FBIS-EEU-94-195* (7 October 1994): 8.

4. Budapest *Magyarorszag* (19 August 1994): 4. *FBIS-EEU-94-178* (14 September 1994): 11.

5. Jan Zielonka perceptively observed that one of the reasons why parliaments and executives have agreed to delegate so much authority to small groups of jurists is because these courts represent for the new political elites a guarantee that their constitutional rights will still be protected even if they find themselves on the losing side of political conflicts. See Jan Zielonka, "New Institutions in the Old East Bloc," *Journal of Democracy,* vol. 5, no. 2 (April 1994): 94–95.

6. *OMRI Daily Report* (7 June 1996): 4–5.

7. *OMRI Daily Report* (20 June 1996): 5.

8. Miklos Haraszti interview, Budapest *Nepszava* (3 August 1997): 1, 7. *FBIS-EEU-96-153* (7 August 1997): 21.

9. Budapest MTI (29 January 1998). *FBIS-EEU-98-029* (30 January 1998).

10. Budapest MTI (26 March 1998). *FBIS-EEU-98-085* (26 March 1998).

11. *Review of Parliamentary Oversight of the Hungarian MOD and Democratic Control of the Hungarian Defence Forces, Study No. 810* (London: Directorate of Management and Consultancy Services, February 1996), 14.

12. Budapest MTI (24 February 1998). *FBIS-EEU-98-056* (2 March 1998).

13. Budapest MTI (18 February 1998). *FBIS-EEU-98-050* (27 February 1998).

14. Defense Minister–designate Gyorgy Keleti interview. *Magyar Nemzet* (13 July 1994): 7. *FBIS-EEU-94-135* (14 July 1994): 13.

15. Colonel Jozsef Feher acted as head of the defense ministry's Administrative and Legal Department until 1992 and then as a staff member of the Institute of Military History.

16. *Heti Vilaggazdasag* (12 November 1994): 103–4. *FBIS-EEU-94-222* (17 November 1994): 22.

17. Budapest Kossuth Radio Network (15 May 1996). *FBIS-EEU-96-096* (16 May 1996): 23.

18. Budapest *Nepszava* (29 March 1995): 1. *FBIS-EEU-95-061* (30 March 1995): 12.

19. *Heti Vilaggazdasag* (12 November 1994): 103–4. *FBIS-EEU-94-222* (17 November 1994): 21–22.

20. Defense Minister–designate Gyorgy Keleti interview. Budapest *Nepszabadsag* (30 June 1994): 1, 4. *FBIS-EEU-94-127* (1 July 1994): 15.

21. Defense Minister Gyorgy Keleti interview. *Magyar Hirlap* (25 July 1994): 3. *FBIS-EEU-94-144* (27 July 1994): 14.

22. Budapest Kossuth Radio Network (7 September 1994). *FBIS-EEU-94-174* (8 September 1994): 13.

23. Budapest MTV (31 October 1994). *FBIS-EEU-94-211* (1 November 1994): 14.

24. *Review of Parliamentary Oversight of the Hungarian MOD and Democratic Control of the Hungarian Defence Forces, Study No. 810* (London: Directorate of Management and Consultancy Services, February 1996), 45–46.

25. Defense Minister Gyorgy Keleti interview. Budapest *Magyar Nemzet* (20 April 1996): 7. *FBIS-EEU-96-079* (23 April 1996): 22.

26. During 1995 the number of conscripts was reduced by five thousand. In February two thousand fewer conscripts were called up, and in August, three thousand fewer. Budapest Kossuth Radio Network (29 June 1995). *FBIS-EEU-95-126* (30 June 1995): 21.

27. Budapest Kossuth Radio Network (21 January 1995). *FBIS-EEU-95-014* (23 January 1995): 27. On 31 January 1995 eighteen thousand soldiers were discharged from service after having served only 11.5 months. Budapest MTV (31 January 1995). *FBIS-EEU-95-021* (1 February 1995): 12.

28. Budapest MTV (13 July 1994). *FBIS-EEU-94-135* (14 July 1994): 13; and Budapest Kossuth Radio Network (20 August 1994). *FBIS-EEU-94-162* (22 August 1994): 15.

29. Budapest Kossuth Radio Network (27 July 1994). *FBIS-EEU-94-145* (28 July 1994): 8.

30. Budapest *Nepszabadsag* (3 August 1994): 1, 4. *FBIS-EEU-94-150* (4 August 1994): 11.

31. Defense Minister Gyorgy Keleti interview. *Nepszabadsag* (8 September 1994): 6. *FBIS-EEU-94-175* (9 September 1994): 16.

32. Budapest MTI (2 January 1995). *FBIS-EEU-95-001* (3 January 1995): 9.

33. Peter Deak, "Armed Forces' Reform," Budapest *Figyelo* (10 August 1995): 17–18. *FBIS-EEU-95-194* (6 October 1995): 21.

34. Groot-Bijgaarden *De Standard* (14–15 May 1995): 4. *FBIS-EEU-95-094* (16 May 1995): 25–26. The Border Guard was to receive sixty BTR-80s and the army forty.

35. Budapest *Magyar Hirlap* (7 February 1996): 1, 4. *FBIS-EEU-96-027* (8 February 1996): 17.

36. Budapest MTV Television Network (29 August 1994). *FBIS-EEU-94-168* (30 August 1994): 18. Keleti cleverly proposed to Ukraine that Hungary would be willing to destroy some of their military vehicles prescribed to Ukraine by CFE. Ukraine would give these vehicles to Hungary who would then destroy its less modern vehicles. Budapest *Magyar Nemzet* (5 January 1995): 6. *FBIS-EEU-95-006* (10 January 1995): 15. On 13 December 1995 Keleti acquired agreement with Defense Minister Valeriy Shmarov. Budapest MTI (13 December 1995).

37. One article criticized Defense Minister Keleti for not consulting the Parliamentary Defense Committee. It argued that Keleti should apologize to the Parliamentary Defense Committee and the committee should thank the minister. See Budapest *Magyar Hirlap* (15 April 1996): 7. *FBIS-EEU-96-074* (16 April 1996): 19.

38. Budapest *Nepszabadsag* (30 March 1996): 4. *FBIS-EEU-96-064* (2 April 1996): 20.

39. Budapest Kossuth Radio Network (9 November 1994). *FBIS-EEU-94-218* (10 November 1994): 18.

40. Budapest MTI (2 May 1995). *FBIS-EEU-95-085* (3 May 1995): 18; Munich *Focus* (19 June 1995): 11. In addition, Germany had trained 118 Hungarian soldiers and delivered 300 wagon-loads of military spares and equipment. *FBIS-EEU-95-151* (7 August 1995): 18.

41. Budapest *Nepszabadsag* (9 November 1995): 27. *FBIS-EEU-95-218* (13 November 1995): 27.

42. Budapest MTI (6 January 1995). *FBIS-EEU-95-005* (9 January 1995): 22.
43. Budapest MTI (1 March 1995). *FBIS-EEU-95-044* (7 March 1995): 8. Also Gyorgy Keleti interview, Budapest *Magyar Honved* (24 February 1995): 6–7. *FBIS-EEU-95-060* (29 March 1995): 23–24.
44. Budapest MTI (22 March 1995). *FBIS-EEU-95-057* (24 March 1995): 8.
45. *Reform of the Armed Forces: 1995–1998–2005* (Budapest: Ministry of Defense, 1995), 4.
46. Budapest *Magyar Honved* (15 September 1995): 7. *FBIS-EEU-95-224* (21 November 1995): 20. Also see *Reform of the Armed Forces: 1995–1998–2005*, pp. 12–13.
47. LTG Sandor Nemeth interview, Budapest *Magyar Hirlap* (7 December 1995): 9. *FBIS-EEU-95-236* (8 December 1995): 14–15; and see *Reform of the Armed Forces: 1995–1998–2005*, p. 11.
48. Defense Minister Gyorgy Keleti interview. Budapest *Nepszava* (8 March 1995): 1, 8. *FBIS-EEU-95-047* (10 March 1995): 12.
49. BG Attila Kositzky interview, Budapest *Magyar Hirlap* (8 December 1995): 6. *FBIS-EEU-95-239* (13 December 1995): 13.
50. Defense Minister Gyorgy Keleti interview, Budapest *Nepszava* (18 September 1995): 3. *FBIS-EEU-95-183* (21 September 1995): 16.
51. Budapest Kossuth Radio Network (30 November 1995). *FBIS-EEU-95-230* (30 November 1995): 17.
52. Budapest Kossuth Radio Network (24 April 1996). *FBIS-EEU-96-081* (25 April 1996): 11.
53. Defense Minister Gyorgy Keleti interview, *Heti Vilaggazdasag* (1 April 1995): 47–48. *FBIS-EEU-95-064* (4 April 1995): 25.
54. Budapest Kossuth Radio Network (4 April 1995). *FBIS-EEU-95-065* (5 April 1995): 30; Defense Minister Keleti interview, *Nepszava* (19 April 1995): 12. *FBIS-EEU-95-076* (20 April 1995): 13.
55. Budapest Kossuth Radio Network (10 May 1995). *FBIS-EEU-95-090* (10 May 1995): 23.
56. *National Defense '95* (Budapest: Ministry of Defense, 14 July 1995), 13.
57. Budapest *Magyar Honved* (15 September 1995): 7. *FBIS-EEU-95-224* (21 November 1995): 20.
58. Budapest *Magyar Hirlap* (15 May 1995): 6. *FBIS-EEU-95-094* (16 May 1995): 29–30; Budapest MTI (27 June 1995). *FBIS-EEU-95-126* (30 June 1995): 7.
59. Budapest *Magyar Hirlap* (4 May 1995): 4. *FBIS-EEU-95-087* (5 May 1995): 5.
60. Budapest *Nepszabadsag* (8 May 1996): 1, 5. *FBIS-EU-96-091* (9 May 1996): 38–39.
61. Budapest *Magyar Honved* (4 August 1995): 6. *FBIS-EEU-95-187* (27 September 1995): 14.
62. Budapest Kossuth Radio Network (12 April 1996). *FBIS-EEU-96-073* (12 April 1996): 19.
63. Budapest Kossuth Radio (5 May 1996). *FBIS-EEU-96-089* (7 May 1996): 21; Budapest Kossuth Radio Network (16 May 1996). *FBIS-EEU-96-096* (16 May 1996): 24.

64. Budapest MTI (22 April 1996). *FBIS-EEU-96-084* (30 April 1996): 4.

65. Budapest *Reform* (2 April 1996): 29. *FBIS-EEU-96-068* (8 April 1996): 10.

66. *Magyar Hirlap* (17 March 1995): 8. Cited in Rudolf Joo, *The Democratic Control of the Armed Forces*, Chaillot Papers 23 (Paris: Institute for Security Studies, WEU, February 1996), 47.

67. Defense Minister Gyorgy Keleti interview, *Heti Vilaggazdasag* (1 April 1995): 47–48.

68. Jozsef Feher interview, Budapest *Nepszabadsag* (20 May 1995): 1, 8. *FBIS-EEU-95-100* (24 May 1995): 18.

69. Budapest *Magyar Hirlap* (12 June 1996): 9. *FBIS-EEU-96-115* (13 June 1996): 27.

70. "Behind Declarations: Civil Military Relations in Central Europe," *Defence Studies: Special Edition* (Budapest: Institute For Strategic and Defence Studies, 1996), 18.

71. Budapest Kossuth Radio Network (18 April 1996). *FBIS-EEU-96-077* (19 April 1996): 19.

72. Budapest Kossuth Radio Network (21 May 1996). *FBIS-EEU-96-100* (22 May 1996): 19; *Nepszabadsag* (29 May 1996): 4. *FBIS-EEU-96-105* (30 May 1996): 17–18.

73. Imre Mecs interview, Budapest *Magyar Nemzet* (29 June 1996): 6. *FBIS-EEU-96-128* (2 July 1996): 15.

74. Budapest *Nepszava* (4 July 1996): 2. *FBIS-EEU-96-131* (8 July 1996): 29.

75. Thomas Wachsler personal interview (25 June 1996).

76. Budapest Kossuth Radio (31 December 1996). *FBIS-EEU-97-001* (3 January 1997).

77. Zoltan Martinusz interview, Budapest *Nepszava* (25 June 1997): 12. *FBIS-EEU-97-199* (18 July 1997).

78. *Magyar Nemzet* (31 August 1995): 6. *FBIS-EEU-95-170* (1 September 1995): 16–17; Budapest Kossuth Radio (19 September 1995). *FBIS-EEU-95-182* (20 September 1995): 20. Two radars would be set up on the summits of the Bukk and Mecsek Mountains to detect aircraft at altitudes between a few hundred meters and 30,000 meters with a range of 400–500 kilometers and a digital communications link to pass signals automatically to the control center and air defense. The new radar uses less energy and would reduce the requirement for a few thousand soldiers. Budapest *Magyar Hirlap* (3 October 1995): 5. *FBIS-EEU-95-193* (5 October 1995): 24.

79. Budapest Kossuth Radio Network (1 November 1995). *FBIS-EEU-95-212* (2 November 1995): 9.

80. Defense Minister Gyorgy Keleti interview (21 September 1995). *FBIS-EEU-95-184* (22 September 1995): 16.

81. Budapest MTV Television Network (9 April 1996). *FBIS-EEU-96-071* (9 April 1996): 15.

82. Defense Minister Gyorgy Keleti interview, Budapest *Magyar Hirlap* (7 May 1996): 10. *FBIS-EEU-96-090* (8 May 1996): 18.

83. Col. Gen. Janos Deak interview, Budapest *Magyar Nemzet* (11 November 1995): 7. *FBIS-EEU-95-219* (14 November 1995): 24.

84. Budapest *Magyar Nemzet* (6 February 1996): 1, 3. *FBIS-EEU-96-026* (7 February 1996): 17.

85. *OMRI Daily Digest* (17 May 1996): 4–5.

86. Budapest *Nepszabadsag* (10 May 1996): 5. *FBIS-EEU-96-094* (14 May 1996): 21–22; *OMRI Daily Report* (31 May 1996): 4.

87. Defense Minister Gyorgy Keleti interview, *Magyar Nemzet* (20 April 1996): 7. *FBIS-EEU-96-079* (23 April 1996): 20; and *Magyar Hirlap* (7 May 1996): 10.

88. Budapest *Nepszabadsag* (2 December 1995): 4. *FBIS-EEU-95-234* (6 December 1995): 21.

89. Defense Minister Keleti interview, Budapest *Magyar Hirlap* (27 October 1995): 1. *FBIS-EEU-95-211* (1 November 1995): 18–19.

90. Budapest *Magyar Honved* (8 December 1995): 6–7. *FBIS-EU-96-030* (13 February 1996): 11.

91. Budapest MTI (9 February 1996). *FBIS-EEU-96-033* (16 February 1996): 8.

92. Budapest Kossuth Radio Network (30 May 1996). *FBIS-EEU-96-105* (30 May 1996): 18.

93. MG Ferenc Vegh interview, Budapest *Nepszava* (11 May 1996): 1, 8. *FBIS-EEU-96-098* (20 May 1996): 16.

94. Budapest Kossuth Radio Network (6 June 1996). *FBIS-EEU-96-111* (7 June 1996): 22.

95. Ferenc Vegh interview, Budapest *Magyar Nemzet* (15 June 1996): 6. *FBIS-EEU-96-118* (18 June 1996): 22.

96. Budapest Kossuth Radio Network (8 June 1996). *FBIS-EEU-96-112* (10 June 1996): 28.

97. Budapest Kossuth Radio Network (31 March 1995). *FBIS-EEU-95-063* (3 April 1995): 11; Budapest Kossuth Radio Network (10 May 1995). *FBIS-EEU-95-091* (11 May 1995): 25.

98. Budapest MTI (1 May 1996). *FBIS-EEU-96-087* (3 May 1996): 13.

99. Imre Mecs interview, Budapest *Nepszabadsag* (21 January 1995): 1, 7. *FBIS-EEU-95-015* (24 January 1995): 36. On 1 September 1995 a language training center was scheduled to open in Budapest. Also Ft12–15 billion would be necessary over many years to modernize the army's telecommunications center. Budapest MTV Television (14 February 1995). *FBIS-EEU-95-031* (15 February 1995): 9.

100. MG Lajos Urban interview, *Magyar Nemzet* (6 October 1995): 8. *FBIS-EEU-95-196* (11 October 1995): 12.

101. Budapest MTI (7 June 1996). *FBIS-EEU-96-114* (12 June 1996): 10.

102. Defense Minister Gyorgy Keleti interview, Kossuth Radio Network (13 August 1995). *FBIS-EEU-95-156* (14 August 1995): 14; Budapest Kossuth Radio Network (1 September 1995). *FBIS-EEU-95-171* (5 September 1995): 19.

103. Budapest Kossuth Radio Network (18 November 1995). *FBIS-EEU-95-223* (20 November 1995): 20.

104. Defense Minister Gyorgy Keleti interview, Budapest *Nepszava* (5 December 1995): 1, 3. *FBIS-EEU-95-234* (6 December 1995): 17–18.

105. Vienna *Kurier* (6 May 1996): 4. *FBIS-EU-96-089* (7 May 1996): 21.

106. Budapest MTI (27 March 1996). *FBIS-EEU-96-064* (2 April 1996): 13.

107. Budapest Kossuth Radio Network (9 April 1996). *FBIS-EEU-96-070* (10 April 1996): 22.

108. Budapest Kossuth Radio Network (3 April 1996). *FBIS-EU-96-065* (3 April 1996): 8; Budapest Kossuth Radio Network (28 April 1996). *FBIS-EEU-96-083* (29 April 1996): 18.

109. Budapest Kossuth Radio Network (9 May 1996); and *Nepszabadsag* (9 May 1996): 20. *FBIS-EEU-96-092* (10 May 1996): 19–20.

110. Budapest Kossuth Radio Network (16 December 1995). *FBIS-EEU-95-244* (20 December 1995): 15.

111. Budapest Kossuth Radio Network (29 February 1996). *FBIS-EEU-96-042* (1 March 1996): 14.

112. Budapest MTI (23 December 1997). *FBIS-EEU-97-357* (29 December 1997).

113. Paris AFP (31 December 1997). *FBIS-EEU-97-365* (2 January 1998).

114. Budapest Kossuth Radio Network (13 May 1996). *FBIS-EEU-96-093* (13 May 1996): 25.

115. Budapest MTI (18 April 1996). *FBIS-EEU-96-080* (24 April 1996): 16.

116. Budapest *Uj Magyarorszag* (10 May 1996): 1, 3. *FBIS-EEU-96-094* (14 May 1996): 21.

117. Budapest MTI (23 April 1996). *FBIS-EEU-96-084* (30 April 1996): 5.

118. Budapest *Magyar Nemzet* (6 May 1996): 1, 4. *FBIS-EEU-96-089* (7 May 1996): 21–22.

119. Budapest *Nepszava* (20 May 1996): 1, 3. *FBIS-EEU-96-100* (22 May 1996): 19.

120. Retired General Bela Gyuricza interview, Budapest *Nepszabadsag* (29 July 1996): 1, 11. *FBIS-EEU-96-147* (30 July 1996): 20.

121. Budapest *Nepszabadsag* (13 August 1996): 1, 5. *FBIS-EEU-96-158* (14 August 1996): 18.

122. Defense Minister Gyorgy Keleti interview, Budapest *Magyar Nemzet* (15 August 1996): 1, 4. *FBIS-EEU-96-162* (20 August 1996): 13.

123. Budapest *Nepszabadsag* (3 August 1996): 1, 4. *FBIS-EEU-96-152* (6 August 1996): 15–16.

124. Budapest MTI (7 November 1996). *FBIS-EEU-96-219* (7 November 1996); Budapest Kossuth Radio (12 December 1996). *FBIS-EEU-96-241* (12 December 1996); and Budapest Kossuth Radio (16 May 1997). *FBIS-EEU-97-136* (16 May 1997).

125. LTG Ferenc Vegh interview, Budapest *Nepszabadsag* (8 October 1996): 1, 8. *FBIS-EEU-96-197* (8 October 1996).

126. Budapest *Nepszabadsag* (28 May 1997): 1,4. *FBIS-EEU-97-149* (29 May 1997).

127. Budapest *Nepszabadsag* (19 June 1997): 4. *FBIS-EEU-97-170* (20 June 1997).

128. Budapest MTI (11 December 1997). *FBIS-EEU-97-345* (12 December 1997).

129. Budapest *Nepszava* (13 December 1996): 1–2. *FBIS-EEU-96-242* (13 December 1996).

130. Defense Minister Gyorgy Keleti interview, Budapest *Nepszava* (8 March 1997): 1, 8. *FBIS-EEU-97-070* (11 March 1997).

131. Budapest *Nepszava* (8 October 1996): 4. *FBIS-EEU-96-197* (8 October 1996).

132. Defense Minister Gyorgy Keleti interview, Budapest *Nepszabadsag* (21 May 1997): 7. *FBIS-EEU-97-099* (23 May 1997).

133. General Jozsef Hollo interview, Budapest Kossuth Radio (4 March 1997). *FBIS-EEU-97-063* (4 March 1997).

134. Budapest *Magyar Nemzet* (8 May 1997): 4. *FBIS-EEU-97-129* (9 May 1997); LTG Ferenc Vegh interview, Budapest *Magyar Nemzet* (5 May 1997). *FBIS-EEU-97-087* (5 May 1997).

135. Budapest *Tallozo* (22 May 1997): 927. *FBIS-EEU-97-113* (13 June 1997).

136. Budapest MTI (19 September 1996). *FBIS-EEU-96-185* (19 September 1996).

137. Budapest Kossuth Radio (5 November 1997). *FBIS-EEU-97-309* (6 November 1997); and MTI (10 April 1998). *FBIS-EEU-98-100* (13 April 1998).

138. Defense Minister Gyorgy Keleti interview, Budapest Kossuth Radio (5 February 1997). *FBIS-EEU-97-025* (5 February 1997).

139. Of the original nineteen MiG-23s purchased in 1979, five crashed over the years. Of the original fifteen Su-22s, twelve survived. Hungarian pilots began flying MiG-21s in 1971. The younger vintage will be retained through 2000. Budapest Kossuth Radio (28 March 1997). *FBIS-EEU-97-087* (28 March 1997).

140. Defense Minister Gyorgy Keleti interview, Budapest Kossuth Radio (3 April 1997). *FBIS-EEU-97-093* (3 April 1997).

141. LTG Nandor Hollosi interview, Budapest *Nepszava* (4 April 1997): 2. *FBIS-EEU-97-097* (7 April 1997).

142. Budapest *Nepszava* (14 March 1998): 3. *FBIS-EEU-98-075* (17 March 1998).

143. Defense Minister Gyorgy Keleti interview, Budapest Kossuth Radio (3 April 1997). *FBIS-EEU-97-093* (3 April 1997).

144. Budapest *Democrata* (13 March 1997): 18–20. *FBIS-EEU-97-079* (24 April 1997). Budapest *Nepszava* (28 June 1997): 2. *FBIS-EEU-97-182* (1 July 1997).

145. "The Russians Again Offer MiGs but Have Little Chance," Budapest *Nepszabadsag* (26 April 1997): 1, 5. *FBIS-EEU-97-118* (28 April 1997).

146. Budapest MTI (31 October 1996). *FBIS-EEU-97-216* (31 October 1996).

147. Budapest *Magyar Nemzet* (17 June 1997): 12. *FBIS-EEU-97-134* (16 July 1997); Budapest MTI (3 December 1997). *FBIS-EEU-97-337* (4 December 1997).

148. Budapest *Nepszabadsag* (4 June 1997): 1, 4. *FBIS-EEU-97-155* (5 June 1977); Budapest *Magyar Nemzet* (24 June 1997): 1, 4. *FBIS-EEU-97-175* (25 June 1997).

149. Budapest *Nepszava* (27 February 1997): 1, 3. *FBIS-EEU-97-040* (3 March 1997); Budapest MTI (19 May 1998). *FBIS-EEU-98-139* (21 May 1998).

150. Budapest *Nepszava* (21 January 1998): 6. *FBIS-EEU-98-021* (23 January 1998).

151. See "The Cost of NATO Enlargement," Atlantica, no. 1 (15 April 1997): 2.

152. Budapest MTI (21 March 1997). *FBIS-EEU-97-083* (21 March 1997).

153. Budapest MTI (10 July 1997). *FBIS-EEU-97-191* (10 July 1997).

154. Colonel Laszlo Nagy, "Thoughts about the Expected Cost of NATO Accession," *Uj Honvegsegi* (1 June 1997): 53–57. *FBIS-EEU-97-153* (11 August 1997).

155. *Opinion Analysis*, M-76-97, p. 1.

156. *Opinion Analysis*, M-66-97, pp. 1–2. On this issue, the Czech Republic is similar to Hungary. Eighty percent of the Czechs believe that military spending increases will accompany membership, and two-thirds would *oppose* such an increase. Eighty percent are opposed when defense spending is pitted against spending for social needs. *Opinion Analysis*, M-79-97, pp. 2–3.

157. The Proceedings of the National Assembly on the Basic Principles of Security Policy in the Parliament of Hungary, galleys 24210-1. Cited in Pal Dunay, "Theological Debates on NATO in Hungary," *Kulpolitika (Foreign Policy)*, vol. 3, 1997 Special Issue (Budapest): 91.

158. Budapest Kossuth Radio (25 August 1997). *FBIS-EEU-97-237* (25 August 1997).

159. Budapest MTI (28 August 1997). *FBIS-EEU-97-240* (28 August 1997).

160. The failed 23–24 May 1997 referendum in Slovakia was designed to elicit precisely the reverse popular response. It asked the following three questions: (1) Do you support NATO membership? (2) Do you support NATO stationing nuclear weapons on Slovak soil? (3) Do you support NATO troops in Slovakia? See Jeffrey Simon, "Slovakia and NATO: The Madrid Summit and After," *Strategic Forum*, no. 111 (April 1997): 2.

161. Budapest MTI (28 August 1997). *FBIS-EEU-97-240* (28 August 1997).

162. See Sean Kay and Hans Binnendijk, "After the Madrid Summit: Parliamentary Ratification of NATO Enlargement," *Strategic Forum*, no. 107 (March 1997): 1–4.

163. Data provided by Hungarian defense attache in Washington, D.C., on 2 September 1997.

164. Gyula Horn news conference, Budapest DUNA (16 November 1997). *FBIS-EEU-97-321* (19 November 1997).

165. LTG Ferenc Vegh interview, Budapest *Magyar Hirlap* (19 July 1997): 8. *FBIS-EEU-97-203* (22 July 1997).

166. Ibid.

167. LTG Ferenc Vegh, Budapest MTI (4 November 1997). *FBIS-EEU-97-308* (6 November 1997).

168. Budapest MTI (1 September 1997). *FBIS-EEU-97-244* (1 September 1997).

169. Budapest Kossuth Radio Network (15 July 1997). *FBIS-EEU-97-196* (16 July 1997).

170. Training commander of the Hungarian Army BG Tamas Suetoe interview, Budapest Kossuth Radio (3 November 1997). *FBIS-EEU-97-308* (6 November 1997).

171. Budapest Kossuth Radio (9 February 1998). *FBIS-EEU-98-040* (10 February 1998).

172. Budapest MTI (25 February 1998). *FBIS-EEU-98-057* (2 March 1998).

173. MG Istvan Szalai personal interview, Budapest (17 April 1998).

174. Jeno Poda interview, Budapest *Magyar Nemzet* (8 December 1997): 4. *FBIS-EEU-97-344* (11 December 1997).

175. Budapest *Nepszabadsag* (27 February 1998): 5.

176. Budapest *Nepszabadsag* (11 August 1997): 4. *FBIS-EEU-97-220* (13 August 1997).

177. MG Istvan Szalai personal interview (17 April 1998).

178. Budapest Kossuth Radio (5 August 1997). *FBIS-EEU-97-217* (5 August 1997); Budapest MTV (2 October 1997). *FBIS-EEU-97-276* (3 October 1997).

179. Budapest *Nepszava* (16 October 1997): 2. *FBIS-EEU-97-289* (20 October 1997).

180. Budapest MTI (2 June 1998). *FBIS-EEU-98-153* (3 June 1998).

181. See Jeffrey Simon, "The IFOR/SFOR Experience: Lessons Learned by PFP Partners," *Strategic Forum*, no. 120 (July 1997): 1–2.

182. Budapest MTI (1 April 1998). *FBIS-EEU-98-091* (3 April 1998).

183. On 20 March 1998 Defense Ministers Keleti and Constantin DuDu Ionescu signed the agreement.

184. Budapest MTI (28 October 1997). *FBIS-EEU-97-301* (28 October 1997).

185. Budapest *Nepszava* (25 June 1997): 12. *FBIS-EEU-97-199* (18 July 1997).

186. Budapest MTI (24 July 1997). *FBIS-EEU-97-205* (25 July 1997).

187. Budapest Kossuth Radio (6 February 1998). *FBIS-EEU-98-040* (10 February 1998).

188. Budapest Kossuth Radio (9 March 1998). *FBIS-EEU-98-068* (11 March 1998).

189. Budapest MTI (17 March 1998). *FBIS-EEU-98-076* (19 March 1998).

190. Deputy State Secretary for Defense Istvan Gyarmati interview, Budapest *Uj Magyarorszag* (6 November 1997): 3. *FBIS-EEU-97-310* (7 November 1997).

191. Budapest MTI (26 October 1997). *FBIS-EEU-97-299* (26 October 1997).

192. Budapest *Nepszabadsag* (12 September 1997): 3. *FBIS-EEU-97-255* (15 September 1997).

193. Budapest MTI (18 February 1998). *FBIS-EEU-98-049* (27 February 1998).

194. LTG Ferenc Vegh interview, Budapest *Magyar Hirlap* (20 November 1997): 3. *FBIS-EEU-97-324* (23 November 1997).

195. Budapest Kossuth Radio (1 February 1998). *FBIS-EEU-98-032* (3 February 1998).

196. Budapest MTI (18 February 1998). *FBIS-EEU-98-050* (27 February 1998).

197. Budapest MTI (2 December 1997). *FBIS-EEU-97-337* (4 December 1997).

198. Budapest MTI (13 January 1998). *FBIS-EEU-98-014* (15 January 1998).

199. Budapest *Magyar Honved* (3 April 1998): 14–19. *FBIS-EEU-98-114* (27 April 1998).

200. Col. Jozsef Varga (ret.), "A Word about Language Learning," *Magyar Honved* (17 April 1998): 33–34.

201. Budapest Duna TV (10 February 1998). *FBIS-EEU-98-041* (12 February 1998).
202. Budapest MTI (17 February 1998). *FBIS-EEU-98-048* (27 February 1998).
203. Budapest MTI (18 February 1998). *FBIS-EEU-98-049* (27 February 1998).
204. Budapest MTI (20 February 1998). *FBIS-EEU-98-051* (1 March 1998).
205. Budapest *Nepszava* (2 March 1998): 1, 6. *FBIS-EEU-98-063* (19 March 1998).

CHAPTER 5

1. Budapest MTI (25 June 1998). *FBIS-EEU-98-176* (26 June 1998).
2. Budapest Kossuth Radio (11 November 1998).
3. Budapest *Magyar Hirlap* (30 June 1998): 4. *FBIS-EEU-98-181* (1 July 1998).
4. Defense Minister Janos Szabo interview, Budapest *168 Ora* (7 July 1998). *FBIS-EEU-98-205* (27 July 1998).
5. Budapest Kossuth Radio (9 July 1998). *FBIS-EEU-98-191* (13 July 1998).
6. Defense Minister Janos Szabo interview, *Magyar Nemzet* (18 July 1998): 7; and *Magyar Demokrata* (13 August 1998): 30–33. *FBIS-EEU-98-230* (19 August 1998).
7. Budapest Kossuth Radio (17 September 1998).
8. Budapest MTI (27 July 1998). *FBIS-EEU-98-208* (28 July 1998).
9. Budapest *Magyar Hirlap* (27 July 1998): 1, 5. *FBIS-EEU-98-208* (29 July 1998).
10. Budapest *Magyar Nemzet* (29 July 1998). *FBIS-EEU-98-218* (10 August 1998).
11. Budapest *Magyar Hirlap* (4 November 1998): 5.
12. Budapest *Nepszava* (30 November 1998).
13. Budapest *Magyar Hirlap* (12 February 1999): 1, 3.
14. Budapest *Nepszava* (8 October 1998): 3.
15. Budapest MTI (16 September 1998).
16. Budapest *Magyar Hirlap* (17 November 1998): 3.
17. Budapest *Nepszava* (3 July 1998): 3.
18. Zsolt Lanyi, chairman of the Parliament Defense Committee, Budapest *Magyar Nemzet* (6 October 1998): 1, 5.
19. Budapest *Magyar Hirlap* (25 November 1998): 3.
20. Budapest *Magyar Hirlap* (18 November 1998): 1.
21. Budapest MTI (3 December 1998).
22. Budapest Kossuth Radio (7 December 1998).
23. Budapest Kossuth Radio (9 March 1999).
24. Budapest Kossuth Radio (14 January 1999).
25. Budapest *Magyar Nemzet* (19 January 1999): 1, 4.
26. Budapest MTI (24 March 1999).
27. Budapest MTI (17 June 1998). *FBIS-EEU-98-168* (19 June 1998).

28. Budapest MTI (27 November 1998).

29. Budapest Kossuth Radio (26 May and 2 June 1999).

30. Budapest Kossuth Radio (9 February 1999).

31. Col. Gen. Ferenc Vegh interview, Budapest *Magyar Nemzet* (12 March 1999): 4.

32. Budapest MTI (17 March 1999).

33. LTG Lajos Urban interview, Budapest *Nepszava* (20 March 1999): 10.

34. Defense Minister Janos Szabo interview, Budapest *Nepszabadsag* (12 March 1999): 3.

35. Budapest *Napi Magyarorszag* (20 March 1999): 3.

36. Budapest *Nepszabadsag* (20 April 1999): 4. The Russians vigorously denied Janos Szabo's allegation. *Nepszabadsag* (21 April 1999): 1.

37. Budapest Kossuth Radio (24 March 1999).

38. Viktor Orban 25 March 1999 press conference.

39. Budapest MTI (13 April 1999).

40. Budapest MTI (8 April 1999). This leak from the closed session caused an investigation. Budapest *Nepszava* (23 April 1999): 9.

41. Budapest *Nepszava* (23 April 1999): 9.

42. "Roundup of Global Reaction to Kosovo Crisis," *European Opinion Alert*, L-19-1999 (Washington, D.C.: USIA, 16 April 1999), 4.

43. Budapest Kossuth Radio (22 April 1999).

44. Budapest Kossuth Radio (27 April 1999).

45. Budapest MTI (29 April 1999).

46. "No to Unlimited Use of Airspace," Budapest *Magyar Hirlap* (7 May 1999): 3.

47. Budapest Kossuth Radio (4 May 1999).

48. Viktor Orban news conference, Budapest Kossuth Radio (26 April 1999).

49. Budapest *Nepszava* (29 April 1999): 1.

50. Budapest *Magyar Hirlap* (7 May 1999): 3.

51. Foreign Minister Janos Martonyi interview, Prague *Pravo* (10 May 1999): 1, 6.

52. Budapest MTI (25 May 1999).

53. Budapest Kossuth Radio (31 May 1999).

54. Budapest Kossuth Radio (2 June 1999).

55. Budapest MTI (11 May 1999).

56. Budapest Kossuth Radio (9 August 1999).

57. Budapest Kossuth Radio (1 December 1999).

58. Budapest *Heti Vilaggazdasag* (29 May 1999).

59. Budapest MTI (2 November 1999).

60. Budapest *Nepszabadsag* (25 November 1999).

61. Deputy Defense Minister Tamas Wachsler interview, Budapest *Nepszabadsag* (31 May 1999): 5.

62. Budapest MTI (4 June 1999).

63. Budapest Kossuth Radio (4 June 1999).

64. Budapest *Napi Magyarorszag* (16 June 1999): 2; *Magyar Hirlap* (11 June 1999): 3.

65. Col. Gen. Lajos Fodor interview, Budapest *Nepszava* (11 August 1999): 1, 6.
66. Budapest *Nepszabadsag* (17 June 1999): 1, 4.
67. Budapest *Nepszava* (2 July 1999): 1, 3.
68. Budapest MTV (1 July 1999).
69. Budapest MTI (9 July 1999).
70. Budapest Kossuth Radio (13 July 1999).
71. Budapest Kossuth Radio (13 May 1999).
72. Budapest MTI (25 May 1999).
73. Budapest *Napi Magyarorszag* (28 May 1999): 4.
74. Budapest *Nepszava* (14 May 1999): 1, 3.
75. Budapest Kossuth Radio (17 June 1999).
76. Budapest MTV Television (14 June 1999); Kossuth Radio (15 June 1999).
77. Budapest Kossuth Radio (8 July 1999).
78. Budapest *Magyar Nemzet* (16 October 1999): 6.
79. Budapest MTI (21 December 1999).
80. Budapest MTI (23 June 1999).
81. Budapest *Magyar Hirlap* (23 June 1999): 1.
82. Budapest Kossuth Radio (14 September 1999).
83. Deputy State Secretary Janos Szabo interview, Budapest *Nepszava* (8 September 1999): 1, 3.
84. Budapest MTI (12 July 1999).
85. BG Lajos Erdelyi interview, Budapest MTI (9 November 1999); and MTI (24 November 1999).
86. Budapest *Nepszava* (2 July 1999): 1, 3.
87. Budapest MTI (15 July 1999).
88. Budapest MTI (16 July 1999).
89. Budapest *Magyar Hirlap* (21 July 1999): 1.
90. Budapest *Magyar Hirlap* (24 July 1999): 1, 4.
91. Budapest *Magyar Nemzet* (6 August 1999): 1, 4.
92. Ferenc Gazdag, "What Size of Army Do We Need?" *Magyar Nemzet* (5 October 1999): 7.
93. Budapest MTI (27 September 1999).
94. Budapest Kossuth Radio (11 September 1999).
95. "Aircraft Tender Only in 2008," Budapest *Nepszabadsag* (13 October 1999): 1.
96. Budapest *Napi Magyarorszag* (23 November 1999): 1.
97. Defense Minister Janos Szabo interview, Budapest MTI (18 August 1999).
98. Tamas Wachsler brief to Parliament's Foreign Affairs Committee, 24 November 1999. Budapest MTI (24 November 1999).
99. Budapest *Magyar Hirlap* (28 August 1999): 1; and *Nepszabadsag* (8 October 1999): 4.
100. Budapest Kossuth Radio (4 November 1999).
101. "Cutting Army Personnel Costs a Lot of Money," Budapest *Nepszava* (28 September 1999): 3.
102 "Open Letter to the Staff of the Hungarian Army," Budapest *Magyar Honved* (29 October 1999).

103. Budapest *Magyar Hirlap* (8 October 1999); and MTV2 Television Network (20 October 1999).

104. Budapest *Nepszabadsag* (9 November 1999): 5.

105. Budapest *Nepszava* (18 November 1999): 3.

106. Budapest Kossuth Radio (14 December 1999).

107. Deputy Secretary of State for Defense Policy Zoltan Martiniusz interview, Budapest MTI (25 November 1999).

108. "Arms Reduction with No Risk," Budapest *Nepszabadsag* (23 November 1999): 3.

109. "Real Change or Sorting Out?" Budapest *Nepszava* (3 December 1999): 1, 3.

110. Ibid.

111. Budapest MTI (29 December 1999).

112. Budapest *Magyar Nemzet* (5 January 2000): 1, 4.

113. Budapest *Nepszava* (26 January 2000): 3.

114. Budapest *Nepszava* (5 March 2001): 1.

115. Budapest MTI (18 January 2000).

116. Budapest *Nepszabadsag* (24 February 2000): 4.

117. Defense Minister Janos Szabo interview, *Nepszava* (1 April 2000).

118. Budapest Kossuth Radio (15 August 2000); *Nepszabadzag* (16 August 2000): 5.

119. Budapest *Nepszabadsag* (18 October 2000): 1, 5.

120. Budapest *Magyar Nemzet* (4 July 2000): 5.

121. Irene Szabo, "Professional Army Costs Much," *Nepszava* (2 February 2000): 1, 3.

122. Budapest, *Nepszabadsag* (18 October 2000): 1, 5.

123. Irene Szabo, "Army Has Vanished from Under Defense Leadership," *Nepszava* (8 January 2002): 1, 3.

124. Budapest *Nepszava* (6 August 2001): 6.

125. Budapest Radio Kossuth (4 September 2001).

126. Discussion with Brigadier General Lajos Erdelyi, Budapest *Nepszava* (29 October 2001): 3; and Kossuth Radio (21 November 2001).

127. Budapest *Magyar Nemzet* (29 January 2000): 1, 5; MTI (20 March 2000).

128. Budapest *Nepszabadsag* (19 April 2001): 4.

129. Budapest Kossuth Radio (16 August 2001).

130. Budapest *Nepszava* (12 August 2000): 1, 2.

131. Defense Minister Janos Szabo interview, Budapest *Nepszava* (1 April 2000).

132. Budapest *Nepszabadsag* (4 July 2000): 5

133. Budapest MTI (11 July 2000).

134. Budapest *Magyar Nemzet* (31 August 2000): 3.

135. Budapest *Magyar Nemzet* (7 September 2000): 1, 3.

136. Budapest Kossuth Radio (27 December 2000).

137. Budapest MTV2 (1 November 2000).

138. Budapest *Magyar Nemzet* (13 October 2000): 3.

139. Budapest *Nepszava* (31 August 2000): 1, 2.

140. Budapest *Magyar Nemzet* (28 November 2000): 2.

141. Budapest Kossuth Radio (25 October 2000).

142. Budapest Duna TV (27 October 2000).

143. Budapest Kossuth Radio (24 November 2000).

144. Budapest Kossuth Radio (1 November 2000).

145. Budapest MTV2 (30 October 2000); *Nepszabadsag* (31 October 2000): 5.

146. Budapest *Magyar Nemzet* (22 November 2000): 7; *Magyar Hirlap* (22 December 2000): 3.

147. Budapest MTV2 (3 January 2001).

148. Budapest Duna TV (16 January 2001).

149. Budapest *Magyar Nemzet* (9 February 2001): 1, 3.

150. Budapest *Magyar Hirlap* (13 February 2001): 3.

151. Budapest Kossuth Radio (15 February 2001).

152. Budapest *Magyar Nemzet* (20 February 2001): 3.

153. Budapest Kossuth Radio (13 June 2001).

154. Budapest *Nepszabadsag* (19 March 2001): 1, 4.

155. Budapest Kossuth Radio (13 April 2001).

156. Budapest *Nepszava* (23 April 2001): 1, 3.

157. Imre Mecs interview, Budapest *168 Ora* (6 September 2001): 12–13.

158. Budapest MTV2 (19 October 2001); and *Magyar Hirlap* (2 November 2001): 5.

159. Budapest Kossuth Radio (10 September 2001).

160. Budapest MTV (10 September 2001).

161. Budapest Kossuth Radio (16 October 2001).

162. Budapest *Nepszabadsag* (20 November 2001): 5; Kossuth Radio (23 November 2001).

163. Budapest Kossuth Radio (18 December 2001); Budapest MTV2 (20 December 2001).

164. Budapest *Magyar Nemzet* (5 March 2002) (Internet version).

165. Jan Gazdik, "Helicopters against Terrorists," Prague *Mlada Fronta Dnes* (11 February 2002): A4.

166. Personal interview, Budapest, 9 April 2002.

167. Mihaly Bak, "Increasing Interest in Contract Officer Military Service," *Magyar Hirlap* (20 February 2002): 1, 5.

168. Budapest Kossuth Radio (1 May 2002).

169. Imre Czirjak, "Evaluation Conference at the Defense Ministry," *Magyar Nemzet* (8 March 2002): 5.

170. Budapest Kossuth Radio (7 March 2002); Budapest *Nepszava* (8 March 2002): 3.

171. Budapest Kossuth Radio (11 September 2001).

172. Istvan Szent-Ivany, Budapest *Nepszava* (13 September 2001): 1.

173. Budapest MTV2 (24 September 2001).

174. Budapest Kossuth Radio (25 September 2001).

175. Budapest Kossuth Radio (5 October 2001).

176. Foreign Minister Janos Martonyi interview, Budapest MTV2 (4 October 2001).

177. Budapest *Magyar Hirlap* (27 September 2001): 4.

178. Budapest MTV2 (7 October 2001).

179. Budapest MTV2 (17 October 2001).

180. Politics may have played a part in this. The press was speculating that the Fidesz-led government coalition had ceased to have a majority in Parliament because they could no longer rely on all of the FKGP thirty-four votes. This meant they would need to avoid controversies in the run-up to the spring 2002 elections. Budapest *Nepszabadsag* (22 October 2001) (Internet version).

181. Budapest *Nepszabadsag* (2 November 2001): 1, 4.

182. Budapest Kossuth Radio (18 November 2001).

183. Budapest MTV2 (5 December 2001).

184. Budapest Kossuth Radio (20 October 2001).

185. Budapest Duna TV (8 November 2001).

186. Buadpest Kossuth Radio (27 November 2001).

187. Budapest Duna TV (15 December 2001).

188. National Security State Secretary Istvan Simicsko interview, *Magyar Nemzet* (5 January 2002) (Internet version).

189. Mihaly Bak, "New Accents in Defense Strategy," *Magyar Hirlap* (8 November 2001): 1, 5.

190. Mihaly Bak, "National Security Concept Adopted," *Magyar Hirlap* (10 May 2002): 6.

CHAPTER 6

1. Budapest *Magyar Hirlap* (31 January 2002) (Internet version).

2. Budapest Kossuth Radio (12 February 2002).

3. Budapest *Nepszabadsag* (11 April 2002): 6.

4. Budapest Duna TV (10 April 2002).

5. Budapest Kossuth Radio (21 April 2002).

6. Budapest Duna TV (16 May 2002).

7. Ferenc Juhasz interview, Budapest *Magyar Hirlap* (27 May 2002): 6.

8. Ibid.

9. Budapest *Nepszava* (12 June 2002) (Internet version).

10. Budapest Kossuth Radio (28 May 2002 and 6 June 2002).

11. Defense Minister Ferenc Juhasz interview, *Nepszava* (15 June 2002): 6.

12. Budapest Kossuth Radio (22 July 2002).

13. Mihaly Bak, "Searching for a Grip on the Gripen Deal," *Magyar Hirlap* (18 June 2002): 6; Budapest Kossuth Radio (9 July 2002).

14. Budapest MTV2 (22 June 2002).

15. Budapest *Nepszabadzag* (31 July 2002): 4.

16. Budapest Kossuth Radio (30 July 2002).

Index

About the Author

Dr. Jeffrey Simon is senior research fellow in the Institute for National Strategic Studies, National Defense University. Previously he was chief, National Military Strategy Branch at the Strategic Studies Institute, U.S. Army War College. He has taught at Georgetown University, and has held research positions at System Planning Corporation and the RAND Corporation. Dr. Simon's publications include numerous articles and ten books, the most recent being *NATO Enlargement: Opinions and Options* (1995) and *NATO Enlargement and Central Europe: A Study in Civil-Military Relations* (1996). Dr. Simon holds a Ph.D. from the University of Washington and an M.A. from the University of Chicago.